Freer Trade, Sustainability, and the Primary Production Sector in the Southern EU: Unraveling the Evidence from Greece

Economy & Environment

VOLUME 16

The titles published in this series are listed at the end of this volume.

Freer Trade, Sustainability, and the Primary Production Sector in the Southern EU: Unraveling the Evidence from Greece

edited by

Joseph N. Lekakis

Department of Economics,
University of Crete, Greece

KLUWER ACADEMIC PUBLISHERS

DORDRECHT / BOSTON / LONDON

Library of Congress Cataloging-in-Publication Data

ISBN 0-7923-5151-7

Published by Kluwer Academic Publishers,
P.O. Box 17, 3300 AA Dordrecht, The Netherlands.

Sold and distributed in North, Central and South America
by Kluwer Academic Publishers,
101 Philip Drive, Norwell, MA 02061, U.S.A.

In all other countries, sold and distributed
by Kluwer Academic Publishers,
P.O. Box 322, 3300 AH Dordrecht, The Netherlands.

Printed on acid-free paper

Printed in the Netherlands

Contents

Figures

Tables

Contributors

George A. Daoutopoulos, Ph.D.
(Michigan State University)
Professor of Rural Sociology,
Aristotle's University of Thessaloniki,
Thessaloniki

Sophia Efstratoglou, Ph.D.
(North Carolina State University)
Professor of Agricultural Organization & Planning,
Agricultural University of Athens,
Athens

Panos Fousekis, Ph.D.
(Pennsylvania State University)
Researcher of Agricultural Economy,
Institute for Mountainous Agriculture,
Karpenision

Stella Galanopoulou-Sendouca, Ph.D.
(Aristotle's University of Thessaloniki)
Professor of Agronomy,
University of Thessaly,
Volos

Giannis Karagiannis, Ph.D.
(University of Saskatchewan),
Researcher of Agricultural Economy,
National Agricultural Research Foundation,
Athens

Stylianos Katranidis, Ph.D.
(University of Kiel)
Associate Professor of Agricultural Economics,
University of Macedonia,
Thessaloniki

Joseph N. Lekakis, Ph.D.
(The University of Michigan)
Associate Professor of Resource & Environmental Economics,
University of Crete,
Rethimno

Elisavet Nitsi, Ph.D. candidate
(Florida Atlantic University)
Florida Atlantic University,
Boca Raton

Christos Pantzios, Ph.D.
(University of Florida)
Research fellow of Agricultural Economy,
National Agricultural Research Foundation,
Athens

Evangelos Papanagiotou, Ph.D.
(University of Leeds)
Professor of Agricultural and Resource Economics,
Aristotle's University of Thessaloniki,
Thessaloniki

Myrto Pyrovetsi, Ph.D.
(Michigan State University)
Associate Professor of Ecology,
Aristotle's University of Thessaloniki,
Thessaloniki

Nikolaos Stamou, Ph.D.
(University of Freiburg)
Professor of Forest Resource Economics,
Aristotle's University of Thessaloniki,
Thessaloniki

Stavros Theofanides, Ph.D.
(University of Manchester)
Professor of Applied Economics,
Pantion University of Athens,
Athens

Evangelos Tzouvelekas, Ph.D.
(University of Wales),
University of Wales,
Aberystwyth

Alexandra Vakrou, Ph.D.
(University of Aberdeen)
Research fellow of Resource Economics,
National Agricultural Research Foundation,
Athens

George Vaphiades, Ph.D.
(University of Edinburgh)
Senior Agricultural Researcher,
Agricultural Bank of Greece,
Athens

Christos M. Zioganas, Ph.D.
(Wye College, University of London)
Professor of Agricultural Economics,
Aristotle's University of Thessaloniki,
Thessaloniki

Preface

Freer Trade, Sustainability, and the Primary Production Sector in the Southern European Union is the first full-scale academic work to capture the primary production sector policy aspects of trade liberalization and sustainability with a detailed focus on a typical southern EU country, Greece.

Many efforts were pooled together in making this book. In May 1996 the Department of Economics of the University of Crete organized an international conference on European Agriculture in the light of the recent WTO agreement and the need for sustainable development. The conference was sponsored by the European Commission (DG XII, contract no ENV4-CT-96-6514), the Economic and Social Research Council (ESRC, contract no L320263049), the Hellenic Ministries for Agriculture and for the Environment, and the University of Crete. While summarizing the conclusions of the conference, several speakers felt that the issues relating to the Southern EU would deserve separate coverage. The conference directors judged that coverage of the issues for a single Southern EU country might have an advantage, regarding detail, over a cross country analysis which would certainly be welcome at a later stage.

Responding to this call, the University of Crete generously reallocated all of its infra-structural costs covered by DG XII, ESRC, and the other sponsors above, in the form of a new grant to host a follow-up conference aiming to explore the issues for Greece. An invitation to all known experts in the field was issued and the conference finally took place in Rethimno between 17-19 October 1997. Selected papers prepared for the conference were subsequently revised on the basis of panel leaders', discussants', and individual participants' comments. An edited volume proposal package was evaluated by Kluwer Academic Publishers, and final revisions were made prior to submitting the final manuscript for publication.

I am gratefully acknowledging the direct and indirect financial support received from the sponsors of the above conferences, the help of the reviewers of this volume, and the support of Kluwer Academic Publishers' Biosciences Division. Last but not least, help with language and other editorials was provided by David Abler, Margaret Loseby, Melissa Peterson, Tim Phipps, Fiona Smith, and Neil Ward.

Rethimno, 31 March 1998
Joseph N. Lekakis

Introduction

Trade, Sustainability, and the Primary Production Sector: A Southern EU Perspective

Joseph N. Lekakis

In the beginning of the present decade, «trade and environment» emerged as one of the most controversial scientific and policy debates of our time. It is now surpassing its infancy, and is suggesting topics and avenues for related research. Supporters of trade liberalization see trade as a vehicle for increased global welfare and are concerned with the possible use of environmental measures as non-tariff protectionist policies (Low, 1992; Anderson and Blackhurst, 1992; Ford-Runge, 1994). Environmentalists fear the generation of adverse environmental impacts, for freer trade implies, *ceteris paribus*, a more aggressive exploitation of environmental resources (Røpke, 1994). Political and other economists see freer trade as both an agent of decreased welfare and environmental conditions in poor countries (Daly and Goodland, 1994; Harrison, Rutherford, and Tarr, 1997). Finally, law experts place the emphasis of their analyses on the relationship between international agreements intersecting trade and environmental policies (Jackson, 1992; Charnovitz, 1996).

Trade and Sustainability versus Trade and Environment

Concerns over the environmental impacts of free trade in the primary production sector have been voiced mainly from the «North», although several studies have directly or indirectly addressed the problem since the late 1980s. Environmentalists in Canada and the US, for example, are critical of free trade agreements because they force district governments, responsible for forest policies, to abolish tree-replanting subsidies (Schaeffer, 1997). They argue also that the new GATT language could undermine the authority of state and local governments, and the use of wood on a more 'sustainable' basis. Compared to national and supranational governments, local governments are more capable of implementing and adjusting to the various dimensions of sustainability in the social/environmental arena (Maser, 1997). Thus, the «impact of trade on the environment» issue is only part of the yet unexplored «impact of trade on sustainability» issue. In this introductory chapter an attempt is made to delineate those two issues. Subsequently, before presenting the questions addressed by the individual authors in this volume, the chapter highlights some relevant conditions for the Southern EU and for Greece.

Trade and Environment

The environmental impacts of free trade are anticipated in agriculture, forests, and fisheries as follows. In agriculture, the increased transport volumes of agricultural products generates pollution, the exploration of new pathways result in the transport of plant, animal, and insect species, and there will be shifts in regional production patterns as regards scale, composition, and production technology (Ervin, 1996). The shift from food crops to export crops in developing countries is important because production of export crops usually implies enhanced soil nutrient depletion as the natural cycle of soil replenishment is disrupted. Conversely the production of perennials is environmentally beneficial (Heering *et al*, 1996).

To meet the increased demand for timber due to freer international trade, unsustainable forest management may lead not only to a decline in the stock of timber but also to various other environmental impacts including the loss of amenities and forest ecological functions (Barbier, 1994). Environmental impacts may be of a direct or indirect nature. Direct impacts are caused by the removal of trees and damages to their surrounding environment, while indirect ones are due to road building and opening up access to the forest.

Increased fishery trade flows can result in overfishing and stock depletion, endangered species exploitation, incidental catches of non-targeted high value, highly tradable fisheries, and the destruction of marine habitats (Sen, 1994).

Some relatively recent literature on the primary production sector has furnished mixed findings. In Sub-Saharan Africa the shift in agricultural production and forestry towards exports may be environmentally beneficial (Repetto, 1989). For export oriented tree and bush crops which are grown with grasses, in West Africa, soil erosion tends to be three times less than crops aimed at domestic markets, such as maize, sorghum, and yam. In Malawi, adjustments have led to changes in the composition and intensity of production rather than changes in the cultivated acreage. Agricultural intensification, along with soil improving crops, helped absorb a rapidly growing population on less land. In Sudan, however, trade and other reforms led to high rates of deforestation by encouraging woodland clearing for crop cultivation to take advantage of higher producer prices (Stryker *et al*, 1989). Other studies looking at national or global deforestation suggest that trade is only one of the many other factors, such as distorted prices, farmer assistance programs, land tenure systems, and regulations, which lead to deforestation (Reed, 1992; Barbier, 1994).

Finally, from a methodological point of view, most recent research suggests that the impact of freer trade on the environment may be separated into scale, mix, externality, policy, and technology, but it remains largely conceptual and exploratory (OECD, 1994; Abler and Shortle, forthcoming).

Trade and Sustainability

Both as a concept and a policy target, sustainability is highly controversial. For some writers, the concept has been stretched to accommodate almost any unrequired social goal because it is seen as a 'good thing' (O'Riordan and Voisey, 1997). For others, the controversy around it exists as a result of the continued debate on whether there is a conflict between environment and development (Baker *et al*, 1997). For worters who have taken this debate further, sustainability has been caught between two conflicting world views: the *expansionist* and the *ecological* (Taylor, 1992). Notwithstanding these ambiguities, three broad approaches to sustainability are emerging from the relevant literature, *the economic, the ecological, and the social.* They have a distinct meaning when applied to the primary sector of production (Munasinghe and McNeely, 1995; Ikerd *et al*, 1996).

The economic approach focuses on achieving a maximum income flow subject to maintaining the natural and man-made capital stock which generates this income. For the primary production sector, economic sustainability implies maintaining some interrelated features including competitiveness and low production costs, farm profitability and income variability, satisfactory returns on investment, and low financial risks and food costs.

The ecological or environmental view of sustainability emphasizes the achievement of stability in bio-physical systems; more specifically, their ability to adapt to changing conditions and withstand stress. In the primary production sector, efforts to improve surface and ground water quality, soil and water conservation, energy efficiency, biological diversity, and farm and product safety, all work towards ecological sustainability.

The social or socio-cultural approach emphasizes the stability of social and cultural systems through equity, preservation of cultural diversities, and pluralism in the decision-making process. Social sustainability in the primary production sector may be monitored by such criteria as the availability of quality food and fiber, employment opportunities, rural landscapes, self-esteem of farmers, ethics of farm practice, and self perceived quality of life.

Liberalizers address sustainability indirectly. Freer trade increases the level of income, but it may interfere with the economic sustainability of the primary production sector through loss of competitiveness. This may prove especially true for regions where this sector enjoyed a substantial level of protection through export and other subsidies. In addition, although additional environmental measures do have an impact on production costs, liberalizers suggests that environmental pollution control costs are a small portion of production costs. Thus, competitiveness is not generally threatened, although it may be for some countries (Ervin, 1996). Whereas, ecological sustainability is prophesized via new environmental controls and increased demand for environmental quality, liberalizers provide little insight into the socio-cultural aspects of sustainability.

Freer Agricultural Trade and Sustainability in the EU

As the debate around freer trade goes on, its relevance to the primary production sector of European Union (EU) countries deserves particular attention. EU agriculture has a specific identity, the CAP reform of 1992 was the instrument used to mediate the Uruguay Round Agreement (URA) in 1993. The US have already introduced a new «hands-free» agricultural policy, the WTO will hold a series of Rounds, the next one anticipated around the year 1999, and the literature on international trade and sustainability with a focus on EU's primary production sector is scarce.

Historically, after attaining self sufficiency, the European Community realized production surpluses, increasing budgetary costs due to subsidized exports, and environmental decay. This, along with pressures towards freer international trade, led to measures, such as land set-aside programs (similar to the US Acreage Reduction Programs, ARP), the abolition of input subsidies and the reduction of export subsidies, aiming to control production. The impact of these programs is not always complete. For example, set-aside schemes are affected by ineffective monitoring, use of more variable inputs per unit of land and increased land productivity. A "clean" agriculture, however, would certainly imply higher production costs and economic hardships for a considerable percentage of the European population engaged in farming.

The WTO, EU initiatives towards sustainable development, and international environmental agreements have changed the dynamics around European, especially southern European agriculture. With a declining labor force and agricultural land availability, southern European agriculture has to face competition within the EU and eventually at the international level. In addition, it must face not only the impact of legal obligations to the WTO, but also domestic sustainability issues which have been carved by the European Commission's «Fifth Environmental Action Programme, Towards Sustainability».

Southern EU agriculture will necessarily follow the adjustments required for a free international trade regime. Recent research suggests that a hypothetical removal of farmer support policies leading to a fully liberalized trade regime would have a rather negligible impact on world food output. However, the shift in production patterns would be significant : North and Latin America, Australasia, and Africa would experience the highest increases, between 5-20 % above their baseline production levels, while Europe and Japan would experience the highest decreases, 15-50% (Anderson and Strutt, 1996). Environmental pressures in the EU would ease, *ceteris paribus*, since the part of the volume of production anticipated to fall is produced in Europe and is associated with the intense use of agrichemicals and intense livestock farming. But, things may be more complicated than hypothesized or expected. The net decline in the production of any EU member country may be composed of uneven responses across its regions. Also, possible land abandonment forced by a decrease in production may cause ecological imbalances due to loss of invaluable wildlife habitat (Ervin, 1996).

While fully liberalized trade regimes exist within trade blocks such as the EU and NAFTA, their materialization on a global scale appears to be an onerous target. What appears most probable are forms of free trade across trade blocks. Therefore, questions of sustainability arise from initiatives taken within these blocks, by administrations both at the top and at the block-member level.

EU agricultural trade policy as reflected in the 1992 CAP reform, as brought to the negotiating table in Uruguay, is highly protectionist.

The same is true with regard to the policies of other contracting members, as the initial levels of protection (those of the 1986-1988 period) were arbitrarily set much higher than the true ones. Thus, the agreed reductions in those levels do not imply any real changes in protection. The CAP reform also attempted to remedy injustices made to the small farmer from the previous policies of price supports; price supports have been blamed for working in favor of the larger farmer. The environmental impacts of the reforms remain exploratory, although the Commission's «Agenda 2000» identifies some positive and some negative impacts. The positive impacts stem from a more rational use of fertilizers and pesticides in response to price decreases. The negative impacts arise from the encouragement of irrigated crops and intensive livestock farming, as a result of the regionalization of direct payments for cereals, oilseeds, and protein crops and through low feed prices respectively. Recent research also advances similar arguments, while recognizing that the CAP reform may have been one of the many factors which affected the decline in agrichemicals, that agricultural production and environmental impacts are location-specific, and that lower prices may in fact lead farmers to change their cropping patterns towards products which require more intensive production methods (Brouwer and van Berkum, forthcoming). Finally, a sudden cutback in farmer support may not lead to positive environmental change (Potter, forthcoming; Beaumond and Barnett, forthcoming).

In «Agenda 2000» the European Commission has been very careful as regards agricultural sustainability under a forthcoming more liberal trade regime. The major features of this Agenda are price stabilization schemes, continued efforts to switch from price to direct support, in order to correct prior injustices made to the small farmers by price support, and specific support towards the enhancement of rural environmental conditions. These features do not, of course, guarantee that agriculture can continue its current pace without its own initiatives to explore new ways to achieve true sustainability.

The «forestry measures in agriculture» and the «agro-environmental measures» which had been introduced by Regulations 2080/92 and 2078/92 respectively, reflect policy determination to support rural incomes affected by the 1992 CAP reform, and encourage a sustainable forestry sector. The objectives of the aid scheme instituted by Regulation 2080/92 were the improvement of forest resources, countryside management practices compatible with environmental balance, and combating the 'greenhouse effect.' Regulation 2078/92 encourages forestry, along with crop and livestock production: to protect and improve the landscape, natural resources, and genetic diversity, as well as upkeep abandoned farmland and woodland; to avert dangers associated with depopulation in rural areas; and, to adopt land management for public access and leisure activities. The seemingly generous aid scheme to forestry does not appear to have influenced land patterns in favor of forestry. Land owners, fearing negative impacts on their land as capital assets, appear reluctant to commit it to a predefined long term use (Loseby, 1996). In addition, products and services supplied by forests are both private and public goods, and financial aid schemes cannot guarantee the achievement of sustainable forestry.

EU's new Common Fisheries Policy (CFP) features some structural measures (e.g., regulations 2081/93 and 2082/93) aiming to modernize and make the fleet more competitive, remove excess capacity, and support coastal communities dependent on fishing (Agiovlasiti, 1998). To reduce the trade deficit in sea products, aquaculture and sea product processing and marketing are also promoted. Conservation of the fish stocks is high in the spirit of the new CFP, which continues to restrict free access to the coastal waters of other member-countries and features distinct Mediterranean fisheries provisions (e.g. regulation 1626/94). These provisions are dictated by, *inter alia,* the high diversity of the fish mass including over 100 species and that, unlike other EU regions, fisheries are mostly artisan in nature exploited under different social and cultural settings. The new CFP in the Mediterranean bans equipment and practices detrimental to marine ecosystems, sets minimum standards for mesh sizes, fishing tools, and fish size, and establishes protected zones.

Policy Issues for a Southern EU Country

It is true that the era of ample farmer support is virtually over, although the recent CAP reform and «Agenda 2000» will guarantee a relative stable price level for agricultural products and some level of direct support up to the year 2006. This is also voiced in recent research and weekly press editorials by known Greek agricultural economists (Sarris, Mergos, and Sarros,1996; Demoussis, 1998; Zanias, 1998). While the European Commission needs to lean over the problems of its southern region with care, a re-orientation of national agricultural policies towards a sustainable agriculture at low economic and social costs in necessary. But how?

For many years now, the small size of the Greek farm is being blamed as the major problem affecting income, productivity and thus competitiveness (see also Zioganas, this volume). This is not necessarily true because low farm income may simply be the result of an efficiently managed small farm. Simultaneously, increased farm size has proven ecologically destructive through the intensive use of chemicals and the disturbance of wildlife habitats. Productivity increases have been at the forefront of all national agricultural development plans and policies since the 1960s. This target was also approached through public infrastructure which is ecologically unsustainable (Lekakis and Kousis, 1994). In other words, ecological functions limit the roles of intensification via new technologies and public infrastructure as agents of increased farm productivity. Thus, the ability of the farm to survive on its own in a liberalized world constitutes a major policy issue. Some farms may be able to develop their own management strategies for survival, while some others may not. Policymakers need to look into this dimension.

Decisions to invest in agricultural production are subject to various kinds of risk and uncertainty, related to environmental and market conditions. Under a relatively stable price level guaranteed by EU's agricultural trade policies for some years to come, risks are related chiefly to the nature of agricultural production, especially crop production. The role of various environmentally damaging inputs, such as fertilizers and pesticides, as risk-increasing or risk-reducing agents, shifts with alternative policy formulations. A liberalized trade regime, through a gradual abandonment of the CAP, will inevitably lead to higher agricultural business risks and a greater demand for risk management. In this frame, the role of agricultural cooperatives needs reassessment. Agricultural policies need to address all these issues very closely.

The route to ecological sustainability, which is prophesized by the removal of farmer support, is by no means guaranteed. The farmers' supply responses to changing prices are vital in identifying this route. Beyond these, policymakers must develop full information arrays regarding farmer persistence in monocultures and industrial crops, as well as alternative solutions.

As stated earlier in this introduction, the controversy around the meaning of sustainability arises from differing views on the prospects of achieving simultaneous economic and ecological targets. If crop production can be maintained without the current levels of agrichemicals and loss in farm income, this target can certainly be met.

Finally, in the sub-sectors of fisheries and forests, policy issues around freer trade and sustainability combine various elements of harvesting rates, production of substitutes, imports, and resource abandonment.

Overview of the Book

The book contains thirteen chapters which have been allocated into three parts. Part I (chapters 1-8) looks at the dynamics concerning effects on the natural environment, and the survival of the Greek agricultural sector. Part II (chapters 9-11), addresses sustainability issues in the biologically renewable natural resources of fisheries and forests. Part III (chapters 12-13), looks at the future and proposes measures to deal with relevant problems at the national and supranational policy level.

In chapter 1, Christos M. Zioganas presents the evolution of the primary agricultural production in Greece during the last thirty years and specifies the new challenges to be faced for re-orientation in the future under the CAP and WTO changing conditions. The evolution of Greek agriculture is examined and compared in two phases, prior to, and after Greece's accession to the European Community. The chapter furnishes estimates in enterprise yields and product prices and values and explains their structure during these phases. It also provides some basic agricultural comparisons between Greece and other EU countries. Finally, the chapter specifies some new relevant challenges to be considered for re-orientation and further development of agriculture in the future, given its historic evolution and the policy environment prescribed by the CAP reforms and the recent and forthcoming WTO regulations.

In chapter 2, Christos Pantzios and Panos Fousekis examine the role of public infrastructure as a policy measure to increasing farm productivity and thus competitiveness. On the one hand, public infrastructural projects in Greek agriculture have caused severe ecological damages directly by converting wetlands to agricultural land and indirectly via irrigation projects which are a prerequisite to intensive agriculture with the known environmental and public health impacts. On the other hand, all these projects have added to productivity increases during the past three decades or so.

In chapter 3, Sophia Efstratoglou assesses the potential of farms to survive in a freer market regime. She asserts that, under the CAP/WTO changing context, the prospects for the farming sector depend to a large degree on its structures and on the strategies which farm households adopt over time. The focus of her chapter is on the farmers' capacity to adjust to more competitive conditions. The adjustment process is reflected in the strategies farm households adopt over time to exploit new market and policy opportunities and overcome constrains. The identification of farm strategies was carried out through a cluster analysis of changes which occurred in a sample of 625 farm household during the period 1981-1991 in two Greek areas, Korinthia and Fthiotis. The analysis rests mainly on observing changes in farm resource use (land, labor, capital, other input) at the farm level, commitment or removal of resources from farming, and response to modernization incentives and policy measures. Three main strategies are identified: the professionalization, the disengagement and the stable reproduction strategy. In a changing market environment, policy makers should seriously take into account the differences in farm household adjustments strategies.

In chapter 4, George Vaphiades' attention is centered on the evaluation of aggregate effects of risk-uncertainty in Greek agriculture, in a multi-period context prior to the recent WTO agreement, by applying the popular Mean-Variance model to the annual income from a sample of fifteen selected crops. The first stage of the empirical investigation uses factor analysis to estimate a composite parameter representing the trend in certain random variables such as acreage, yield, production costs, output, price, etc, which characterize the dynamic evolution of the selected crops. Then, multivariate analysis is applied to test the correlation of these variables. The empirical results do not support the hypothesis that either the average Greek farmer was risk-neutral, or that crop production choices were described by maximizing an expected utility function defined over crop income. In addition, production costs were the only factor highly correlated with income variance, therefore contributing to income risk Thus, the gradual abandonment of the CAP, which is expected to increase competitive conditions in agriculture, will lead to higher business risk for the Greek farmer and a greater demand for risk management. Policy makers should attempt to convince farmers to focus on their business outcomes and the risks inherent in their production choices in a long term framework.

In chapter 5, Giannis Karagiannis considers the validity of the «intensification hypothesis», which asserts that farm sector price supports leads to greater use of fertilizers and pesticides per unit of land, under conditions of risk and production uncertainty. He shows that by assuming cost minimization and a general form of production uncertainty, stochastic complementarity and increasing partial relative risk aversion (IPRRA), a production subsidy results in an increase of a risk-reducing input (i.e., pesticides), while the corresponding effect on risk-increasing input demand (i.e., fertilizer) is ambiguous. However, increases in price support result in an increase of fertilizer use, as long a production uncertainty has a multiplicative or an additive form. This is more likely to occur for nitrogen and phosphorus. On the other hand, if pesticides are considered as a risk-reducing input, an increase in price support causes pesticide use to increase, while the opposite is true for an increase in the level of the corresponsibility levy. Similar results hold for potassium, which in some empirical studies was found to be risk-reducing. Given that it has been found that in Greek agriculture, the elasticity of fertilizer demand with respect to crop price is, in absolute terms, much greater than the own-price elasticity, fertilizer use has been affected more by crop output prices than fertilizer price. Thus, for Greek agriculture, higher price support seems to have a greater influence on fertilizer use than does a reduction of its price.

In chapter 6, Joseph N. Lekakis addresses the proposition that a fall in producer prices following the elimination of farmer support would lead to a cutback in production and thereby an amelioration of environmental offenses due to the more rational use of agrichemicals. He employs the traditional Nerlovian partial adjustment supply response model for such crops as cotton and corn which are the most soil erosive and chemicals intensive crops produced in Greece. The own elasticities of supply for both cotton and corn have been found to be too small to anticipate any real cutbacks in production as a result of lower producer prices. Policy makers, concerned with the economics-environment interface in agriculture, should pay attention to farmers' supply response for crops whose production is aggressive to the environment.

In chapter 7, Evangelos Papanagiotou deals with the application of chemicals which is a major area of economic and environmental concern. Taking into account the magnitude of fertilizer contribution to the cost and productivity of agriculture, his chapter attempts to estimate some negative effects of a chemical limitation policy, especially for farmers, over time. More precisely, a short review of the methods used, their results and shortcomings are given and then a two output and six input model (including chemicals) has been constructed and estimated for two different periods: 1960-90 and 1973-1996 for the whole Greek agricultural sector. This estimation gave fertilizer elasticities with reference to own price, expected crop prices and yields. Finally, using a simulation procedure, dynamic multipliers have been estimated for the exogenous variables of fertilizer and crop production prices. Policy makers need to assess any possible negative impact of fertilizer reduction on farm incomes.

In chapter 8, Stella Galanopoulou-Sendouca implies that a reduction in the level of inputs used does not necessarily lead to lower farm incomes. She looks at various ways of making the most erosive and high in chemical application crop, cotton both competitive and ecologically sustainable. She examines cultivation approaches to producing cotton at lower production costs, with the least possible use of agrichemicals and other inputs. Among other sources of evidence, experimental data derived from a COSSYM/COMAX model show that, under certain conditions, this is feasible. Before proposing policy options, she evaluates other approaches which affect production costs indirectly.

In chapter 9, Panos Fousekis and Christos Pantzios, study the relationship between harvest and fishing effort which is inherently stochastic. Technological innovations and permanent or temporary changes in the density of fish stock make it impossible to determine *a priori* the harvest which will result from a given level of effort. Because of this, standard econometric approaches, which rely on deterministic trends to capture the effects of technological innovations and density changes on the effort-harvest relationship, may lead to inconsistent estimates. In this chapter, stochastic trends are employed to decompose the performance of the bottom trawler fleet in Greece into scale effects and productivity effects. This is achieved by specifying a variant of the standard harvest-effort model and estimating it by the Kalman Filtering Algorithm. The empirical results based on annual data of the period 1964-1995 suggest that: 1) The fleet of bottom trawlers experiences serious diseconomies of scale with respect to the basic harvesting input which is the Horse Power (HP). Because of this, current policy measures attempting to preserve fish stocks through limitations in HP are likely to have a small impact in total harvesting. 2) the diseconomies of scale in combination with the expansion of HP are responsible for the decrease in the efficiency of vessels during the examined period, ; 3) the productivity effects, reflecting the impact of stochastic technological innovations and permanent changes in the density of the fish stocks, have to a certain extent offset the negative effects associated with the diseconomies of scale, and, finally, 4) the productivity effects have been larger than the scale effects during the most recent years (1984-95). Given that in 1983 the EU initiated its policy for modernization and renovation of professional vessels, one may infer that this policy has worked towards an increase in the Fleet's overall productivity.

In chapter 10, Giannis Karagiannis, *et al* examine aquaculture which, in the last decade, has undergone a rapid development and has become one of the most significant sectors of primary production with an appreciable contribution both to primary sector GDP and to exports. The sector's position and prospects also appear to have great importance at the international level. Even though internationally, and especially in Europe, aquaculture exhibits high development rates, to date, Greece holds the first place in Europe as regards production volume and exports. This chapter examines two important aspects related to the sector. The first refers to its development prospects, in a competitive international environment, as the sector has a definite export based profile. Productivity and economic efficiency, two of the most important elements for the formation of competitiveness, constitute the variables to be analyzed. The second issue is aquaculture's environmental impact. The reason this has become a front line issue, is that it is the only objection that has been raised for the development of aquaculture in Greece and elsewhere. For this reason, the chapter explores an economic valuation framework for any possible environmental impacts. Aquaculture has been a major tool of the new Common Fisheries Policy (CFP) and policymakers should seek to balance the edges of economic and environmental sustainability in this sub-sector.

In chapter 11, Nikolaos Stamou, proposes that the social viability of forest enterprises depends on their functional improvement, the joint production of environmental products, and the acceptance by society of their environmental profile. The Mediterranean, and the Greek forests in particular, are exceptionally multifunctional offering the satisfaction of needs for forest products as well as the creation of conditions for the viability and sustainable development in other productive sectors such as agriculture. Moreover, the forest is a component of the coastal and the broader natural environment and a strong factor of protection against desertification. These attributes make the forest an important source for both consumptive and non-consumptive uses, and for the maintenance of biodiversity and cultural diversity. Greece has been a country which relied heavily on imports of wood and timber products. Freer international trade in timber will make it easier for the Greek economy to satisfy its relevant demand for imports, and will allow Greek forests to fulfill their multifunctional and particularly their ecological role at a greater degree. At the same time, however, the strong competitive environment which is underway will place the attempt to modernize the forestry sector on entirely new foundations. Thus, although policy makers should not worry about the WTO's negative effects on Greek forests, the economic sustainability of those near them is indeed threatened.

In chapter 12, George A. Daoutopoulos and Myrto Pyrovetsi maintain that, solving the present problems faced by Greek agriculture is not an easy task. More competition and open markets are not the panacea to the Greek farm crisis as CAP uniform measures that ignored local conditions did not help localities in the EU periphery. A sustainable path seems to work better under the present structure of Greek agriculture provided that it takes the form of an integrated program aiming at: informing and educating the farm population, restoring old practices with a sustainable character; monitoring the impact of agriculture on the natural environment; modernizing market structures; completely reforming the rural cooperatives into new structures; and, designing a proper framework for risk management.

Finally, in chapter 13, Stavros M. Theophanides, proposes an alternative route to solving EU's (and thus Southern EU's) agricultural problems, including persistent surpluses, rising financial costs, failure to exploit the comparative advantage of some agricultural products, frequent unrest and mobilization of farmers, weak European position on the world famine problem, and persistent uncertainties. These problems stem from the adoption of the 'policy method' while the efficient strategy is a market-oriented 'planning method.' This chapter supports the thesis that the Common Agricultural Policy (CAP) should be replaced with a Common Agricultural Planning and Management (CAPLAM) approach. Among the many ingredients of this approach are its guidance by ecological-geographic comparative advantage, product diversification and quality improvement, the achievement of endo-industrial and agricultural-industrial balances, comprehensives, external relations, the socio-cultural status of the farm household, and ad hoc priorities.

References

Abler, D. G. and J. S. Shortle (forthcoming) «Decomposing the Effects of Trade on the Environment,» in Antle, J. M, J. N. Lekakis, and G. P. Zanias (eds), *Agriculture, Trade, and the Environment: The Impact of Liberalization on Sustainable Development*, Cheltenham: Edward Elgar.

Agiovlasiti, O. 1998. «Implementation of the New Common Fisheries Policy (CFP),» paper presented at the conference, The Common Agricultural Policy of the EU: New Market Trends, MAICH, Chanea, Crete, published by CIHEAM, *Cahier Options Mediterraneenes*, 29:133-142.

Anderson, K. and R. Blackhurst (eds), 1992. *The Greening of World Trade Issues*, London: Harvester Wheatsheaf.

Anderson, K. and A. Strutt, 1996. «On Measuring the Environmental Impact of Agricultural Trade Liberalization,» in M. Bretahl *et al* (eds), *Agriculture, Trade, and the Environment: Discovering and Measuring the Critical Linkages*, Boulder, Colorado: Westview Press.

Baker, S., M. Kousis, D. Richardson and S. Young, 1997. «Sustainable Development in the European Union: Theoretical and Policy Perspectives,» in S. Baker, M. Kousis, D. Richardson, and S. Young (eds), *The Politics of Sustainable Development: Theory, Policy, and Practice Within the EU*, London: Routledge.

Barbier, E. B. 1994. «The Environmental Effects of Trade in the Forestry Sector,»in *The Environmental Effects of Trade*, OECD, Paris.

Beaumond, H-C, and A. Barnett, (forthcoming). «The Opportunities and Risks for the European Rural Environment Associated with the Liberalization of Agricultural Trade,» in M. R. Redclift, J. N. Lekakis, and G. P. Zanias (eds), *Agriculture and World Trade Liberalization: Socio-environmental Perspectives on the Common Agricultural Policy*, Wallinford, UK: CAB International.

Brouwer, F. and S. van Berkum, forthcoming. «Effects of CAP Reform on the Environment in the EU,» in Antle, J. M, J. N. Lekakis, and G. P. Zanias (eds), *Agriculture, Trade, and the Environment: The Impact of Liberalization on Sustainable Development*, Cheltenham, UK: Edward Elgar.

Charnovitz, S. 1996. «Multilateral Environmental Agreements and Trade Rules,» *Environmental Policy and Law*, 26/4: 163-169.

Daly, H. and R. Goodland, 1994. «An Ecological-Economic Assessment of Deregulation of International Trade under GATT,» *Ecological Economics*, 9: 73-92.

Demoussis, M. 1998. «Viocarpet Junction and Vathis Square,» *To Vima*, weekly paper, 8 February, p. D14 (in Greek).

Ervin, D. 1996. «Agriculture, Trade, and the Environment: Anticipating the Policy Challenges,» OECD, Paris (restricted)

Ford-Runge, C. 1994. *Freer Trade, Protected Environment: Balancing Trade Liberalization and Environmental Interests*, New York: Council of Foreign Relations Press.

Harrison, G. W., T. F. Rutherford, and D. G. Tarr, 1997. «Quantifying the Uruguay Round,» *The Economic Journal*, 107: 1405-1430.

Heerink, N., A. Kuyvenhoven, and F. Qu, 1996. «Policy Issues in International Trade and the Environment with Special Reference to Agriculture,» in M. Munasinghe (ed.), Environmental Impacts of Macroeconomic ans Sectoral Policies, Washington DC: The World Bank.

Jackson, J. H. 1992. «World Trade Rules and Environmental Policies: Congruence or Conflict?» *Washington and Lee Review*, 49 (4): 1227-1277.

Ikerd, J., G. Devino, and S. Traiyongwanich, 1996. «Evaluating the Sustainability of Alternative Farming Systems,» *American Journal of Alternative Agriculture*, 11(1): 25-29.

Lekakis, J. N. and M. Kousis, 1994. «Agriculture and the Environment in Greece,» *Progress in Rural Policy and Planning*, 4: 177-207.

Loseby, M. 1996. «Attitudes to Forestry in the EU: A Survey of Approaches by the European Parliament and the European Commission,» in M. Loseby (ed), *Agroforestry and its Impacts on the Environment*, University of Tuscia and National Institute of Agricultural Economics (INEA), Italy.

Low, P. (ed), 1992. *International Trade and the Environment*, Washington D.C., World Bank Discussion Paper # 159.

Maser, C. 1997. *Sustainable Community Development: Principles and Concepts*, Delray Beach, Florida: St. Lucie Press.

Munasinghe, M. and J. McNeely, 1995. «Key Concepts and Terminology of Sustainable Development,» in M. Munasinghe and W. Shearer (eds), *Defining and Measuring Sustainability*, Washington, D.C.: The World Bank.

OECD, 1994. *Methodologies for Environmental and Trade Reviews*, Paris.

O'Riordan, T. and H. Voisey, 1997. *Sustainable Development in Western Europe: Coming to Terms with Agenda 21*, London: Frank Cass.

Potter, C. (forthcoming). «Agricultural Liberalization and the 'Double Dividend',» in M. Redclift, J. N. Lekakis, and G. P. Zanias (eds), *Agriculture and World Trade Liberalization: Socio-environmental Perspectives on the Common Agricultural Policy*, Wallinford, UK: CAB International

Reed, D. (ed) 1992. *Structural Adjustment and the Environment*, Boulder, Colorado: Westview Press.

Repetto, R. 1989. «Economic Incentives for Sustainable Production,» in G. Schramm and J. Watford (eds), *Environmental Management and Economic Development*, Baltimore: The Johns Hopkins University Press.

Røpke, I. 1994. «Trade, Development, and Sustainability - A Critical Assessment of the Free Trade Dogma,» *Ecological Economics*, 9: 13-22.

Sarris, A., G. Mergos, and P. Sarros, 1996. *The Uruguay Round Agreement for International Trade and its Impact on Greek Agriculture*, Foundation of Economic and Industrial Studies, Special Studies No. 25, Athens.

Schaeffer, R. K. 1997. *Understanding Globalization: The Social Consequences of Political, Economic, and Environmental Change*, New York: Rowman & Littlefield Publishers.

Sen, S. 1994. «The Environmental Effects of Trade in the Fisheries Sector,» in *The Environmental Effects of Trade*, OECD, Paris.

Stryker, J. D. *et al*, 1989. «Linkages between Policy Reform and Natural Resources Management in Sub-Saharan Africa,» unpublished paper, Tufts University, Fletcher School and Associates for International Development.

Taylor, D. M. 1992. «Disagreeing on the Basics: Environmental Debates Reflect Competing World Views,» *Alternatives*, 18(3): 26-33.

Zanias, G. 1998. «The 'Price' of High Prices,» *To Vima*, weekly paper, 8 February, p. D14 (in Greek).

Zioganas, C. M. (this volume). «The Organization of Primary Agricultural Production in Greece: Evolution and New Challenges for Re-Orientation.»

Part I
The Agricultural Sector

Chapter 1

The Organization of Primary Agricultural Production in Greece: Evolution and New Challenges for Re-Orientation

Christos M. Zioganas

Introduction

Since the beginning of the 20th century, the Greek agricultural sector has experienced substantial developments. This is demonstrated by, *inter alia*, the changing conditions and organization of primary production, the level of income, and the level of technology. However, serious structural problems still exist and they have not yet been dealt with decisively. At a time when favorable natural environmental advantages have not been exploited adequately, agriculture will continue to play a key role in the future of the Greek economy.

The present organization of Greek agriculture reflects the impact of both national policies and the Common Agricultural Policy (CAP) before and after the country's accession to the European Union (EU). So far, considerable efforts have been made by various researchers to investigate the impact of the CAP on Greek agriculture, from different angles. The most recent works by Georgakopoulos (1988), Demoussis and Sarris (1988), and Zioganas (1993) have shown that Greek agriculture underwent a slight improvement, particularly as regards farm incomes. However, agricultural trade effects seem to have been negative (Mattas, 1990; Sapounas, 1994).

The rapid and drastic changes in policy which occurred recently, both at the European (CAP reform) and the international (GATT/WTO agreement) level, are imposing a need for the adaptation of the agricultural sectors in all EU member-countries.

This chapter is focused firstly on the evolution of primary agriculture in Greece, looking at its structure, production, and development over the period 1966-1994 (almost equally divided before and after Greece's accession to the EU). Secondly, new challenges faced by policymakers and farmers now indicate a need for re-orientation.

The Organization of Primary Agricultural Production

This section looks at farm size and other structural attributes of Greek agriculture, the distributional profile of farms across sub-sectors, and economic development of the sector.

The structure of Greek agriculture

The agricultural sector of Greece has a poor structure. This is mainly attributed to the very large number of small and fragmented farms, a low proportion of irrigated arable land and a high share of economically active members of the population engaged in farming activities. This structure, in conjunction with an insufficient infrastructure, professional training of primarily old farmers, an inefficient organization of the relevant public services, and the system of marketing, acts as an impediment to the development of the sector. Consequently, farm incomes remain relatively low, in comparison with the incomes from other sectors of the Greek economy and with those from agricultural sectors of other EU member-states (Demoussis and Sarris, 1988). Only those small farms which are located in regions with favorable soil, climate conditions and a proper mix of farm enterprises do ensure the effective utilization of available family labor and a satisfactory level of farm income.

At the beginning of the 1990s arable land in Greece was only 29.7% of the total land area. Furthermore, the number of small farms with up to 50 stremmas (1 str. is about 0.1 ha) of cultivated land exceeded 75% of the total farm population and covered little more than a third of the total cultivated area of the country. However, a slight decrease in both their number and area has gradually taken place over the period 1971-1991. At the same time, the larger farms (over 50 str.) had increased in number and land area.

Despite continuous changes over the period 1966-1994, the average figures expressing structure still remain poor. Table 1.1 indicates that the average farm size in 1994 was 43.5 str. of arable land; the fragmentation on farms shows an average of 5.9 parcels of arable land; average amounts of irrigated arable land expressed as a percentage of total arable land was 31.6, and the economically active population engaged in agriculture was 22.9%.

Comparing the two periods (15 years pre-entry and 14 years post-entry), the average annual rates of change show that for all four structural characteristics, Greece has experienced lower improvement after accession to the EU than before. This means that accession has hardly affected the structure of Greek agriculture, as the funds directed to structural programs have not been capable of accelerating any significant changes.

TABLE 1. 1: Some structural characteristics of Greek agriculture

Years	Structural characteristics			
	Average farm size: str. of arable land	Average fragmentation of farms: no. of parcels of arable land	Irrigated arable land: % of total arable land (2)	Economically active population engaged in agriculture(3) (%)
1965	33.0	6.8	11.8	49.0
1980	41.0	6.1	22.7	30.3
1994(1)	43.5	5.9	31.6	22.9
Annual average rate of change (%)				
1966-1980	1.6	-0.7	6.2	-2.5
1981-1994	0.4	-0.2	2.8	-1.7

(1) Unofficial estimates ; (2) Almost stable with an area of 39,200,000 str.
(3) Approximately a population 1,600,00 in 1965 and 750,000 in 1994.
Source: National Statistical Service of Greece (NSSG).

<u>The sectoral profile of farm enterprises</u>

The gradual implementation of the CAP in Greece has favored a great expansion of industrial crops (cotton, tobacco, and sugar-beets) and citrus-trees (mainly orange-trees). This is in comparison to the relatively lower expansion of vegetables, fruit-trees (particularly peach-trees) and olive-trees. This has primarily occurred by the replacement of some annual crops (mainly cereals) and to a lesser degree perennial crops (mainly vineyards). Table 1.2 presents the evolution of the sectoral profile of farm enterprises over the period 1965-1994 totally as well as over the pre-entry (1966-1980) and post-entry (1981-1994) periods separately. At the same time, while the CAP has favored a slight growth in sheep and goat populations, it boosted poultry and pig production. Bovines (especially dairy) continued to decrease before and after accession.

TABLE 1. 2: The sectoral profile of farm enterprises

	Crop area/Livestock units[1] (%)				
Enterprises	1965	1980	1994	Pre-entry change 1966/80 (%)	Post-entry change 1981/94 (%)
A. Crops	100.0	100.0	100.0		
Cereals	43.6	33.3	28.1	-23.6	-15.6
Vegetables	4.6	5.3	5.9	+15.2	+11.3
Industrial crops	9.7	8.4	12.2	-13.4	+45.2
Fodder crops	15.7	22.9	23.1	+45.9	+0.9
Fruit-trees	1.2	2.0	2.2	+66.7	+10.0
Citrus-trees	1.4	1.5	1.9	+ 7.1	+26.7
Olive-trees	15.8	20.4	21.1	+29.1	+3.4
Vineyards	8.0	6.2	5.5	-22.5	-11.3
B. Livestock	100.0	100.0	100.0		
Bovines	34.0	26.3	22.6	-22.6	-14.1
Sheep	32.1	30.1	30.6	-6.2	+1.7
Goats	15.9	17.8	17.9	+11.9	+0.6
Pigs	9.1	14.8	15.8	+62.6	+6.8
Poultry	8.9	11.0	13.1	+23.6	+19.1
Total (l.u.)	2,442.8	2,678.0	2,679.0	+9.6	+0.04

(1) Assembled on the basis of data from National Statistical Service of Greece (NSSG).

A general observation from experience is that a rather unsatisfactory marketing system exists for almost all agricultural products, which up to now has not assisted most of them, even the traditional ones (e.g. wine, fruit, and vegetables), in achieving a remarkable competitiveness on the international market.

Economic development in agriculture

In this section, reference is made to the evolution of productivity (yields) of farm enterprises (crop and livestock) and the prices of their products. Also, a measurement of agricultural development is attempted in terms of GDP growth rates in the primary production sector.

Productivity and prices

The ultimate results of the change in the structure and mix of farm enterprises in the Greek agricultural sector, as described above, can be traced in the information given in Table 1.3.

TABLE 1. 3: Weighted indices of productivity and prices and annual rates of change

	Indices (1965=100)		Average annual rates of change (%)	
Farm Enterprises	1980	1994	Pre-entry	Post-entry
Crops				
Yields per str.	171.69	200.06	4.78	1.18
Product value per str.	129.06	138.23	1.94	0.48
Producer's prices	69.50	63.88	-2.03	-0.58
Livestock				
Yields per unit	155.60	159.16	3.71	0.16
Product value per unit	157.77	156.68	3.85	-0.05
Producer's prices	100.58	97.65	0.04	-0.21
Total				
Yields	166.54	191.58	4.44	1.07
Product value per str.	139.41	146.83	2.63	0.38
Producer's prices	94.05	87.86	-0.40	-0.47

Source: NSSG. Note: Product values at 1970 prices.

Weighted indices of yields, productivity and prices were computed for crops and livestock separately and totally. In addition, their corresponding average annual rates of change before and after accession to the EU are included. Crops and livestock yields after accession continued to increase but to a lesser degree (particularly in livestock), and the overall yields resulted in a slight increase expressed by a positive annual rate of change 1.07% compared with a corresponding figure 4.44% before accession. The overall yield index increase expresses the "production dynamics" of Greek agriculture. With regard to producers' prices, they have slightly fallen on average for all crops and livestock (much more in crops), having resulted in an overall very slight decrease after accession. Of course, there are differences in yields and price indices among enterprises; the prices have been raised to relatively high levels following accession mainly for industrial crops, olive-trees, sheep, and poultry.

The pattern of yields and prices has resulted in a continuous increase of the product value per unit in general but at a lesser degree over the period after accession (0.38% per year) compared with the period before accession (2.63% per year). More specifically, the product value of crops had a slightly more positive change. But livestock experienced a very small negative one within the second period.

Growth

Despite persistent structural problems over the last three decades or so, an overall increase in both total and per worker agricultural gross domestic product (GDP) has taken place (46.1% and 128.8% respectively). The corresponding average annual growth rates were 1.6% and 4.4.%. Thus, GDP per worker has grown at a much higher rate than total GDP. The share of agriculture in GDP produced by the entire economy has decreased significantly; but this merely indicates a relatively faster output growth in the non-agricultural sectors. Another relevant study (Sapounas, 1994) indicates that in Greece the declining share of agricultural labor to total labor force and the growth of agricultural output per capita have been accompanied by increasing labor skills and education level; nevertheless, the «worker effect» on output has been much stronger in the non-agricultural sectors.

The relative indices conceal the poor impact that the CAP exerts on the agricultural economy of Greece. The average annual rates of growth in the pre-entry period were 2.6% in total and 6.5% per employee, while in the post-entry period the relative figures were 0.3% and 1.1% respectively.

So, a remarkable decrease in labor productivity has taken place in the post-entry period in relation to the pre-entry period. The most probable explanation could be that the rate of substituting labor inputs by capital inputs has fallen (investments of fixed capital have been gradually declined, having indices: 100 in 1971, 70.3 in 1981 and 56.5 in 1991, at constant 1970 prices) and consequent unemployment in the non-agricultural sectors has prevented the labor force from leaving agriculture.

However, the development of Greek agriculture has continued after accession, even at relatively low rates. This can be attributed to favorable changes in the farm enterprise mix (mainly expansion of industrial crops, fruit-trees, sheep, poultry and pigs, and reduction of some of the cereals- particularly soft wheat), increasing productivity (yields), rational use of resources, and, in particular, the circulating capital and labor (Zioganas, 1993).

It should be added that the contribution of the main branches to the total agricultural GDP are approximately: crops and livestock 95.0%, fisheries 3.9% and forestry 1.1%, while the ratio in terms of product value between crops and livestock is presently 68/32, with only a relatively small change since 1965 (when that ratio was 73/27).

It is relevant to point out here that, to a great extent, the performance of agricultural products in international markets reflects both marketing effectiveness and the competitiveness (cost and quality) of the particular products. Throughout the period 1961-1993, even though exports have risen, imports (mainly meat and milk) have also risen, but more rapidly especially after 1981. The agricultural trade balance of the country rose from 4.68 billion drachmas in 1981 to 147.39 billion in 1993. As pointed out elsewhere (Sapounas, 1994), over the last two to three decades, Greek agricultural exports have adjusted rather badly to the changing international economic environment.

Therefore, the increasingly negative trade balance of agricultural products over the last 15 years is a problematic change for Greece. This is primarily because the agricultural sector is very important from the viewpoint of the size of the population engaged in agriculture (22.9%) and of GDP (12.5%), as well as the availability of advantageous natural environment conditions for agricultural production.

Up to now, the development of Greek agriculture has relied on the following policy items:

(a) use of more and more new inputs,

(b) implementation of one-sided efforts to increase quantities of products mainly designed to ensure self-sufficiency,

(c) introduction of generalized systematic and intensive technical and economic support to farmers mainly by the Agricultural Extension Service of the Ministry of Agriculture, and

(d) protection of product prices and farm incomes as well as compensation for production surpluses.

Greece and Other EU Countries

In a comparison between Greece and other EU countries, aspects of interest include structural, technological, and developmental issues. Regarding structural characteristics, Greece holds the lowest proportion of agricultural land used in relation to the total area of the country, and has the smallest average farm size in terms of arable land compared with the other EU countries.

Amongst EU countries each one of the five Mediterranean countries has the highest proportion of perennial crops (ranging from 5.0% to 19.8%) in an agricultural land area which altogether makes a total share 80.1%. The same can be said about the share of the number of farms of each of them in the whole of the EU (ranging from 7.4% to 32.2%) and of the sum of them constituting 82.7% of the total farms in the 12 countries. With the exception of France, the Mediterranean countries have the smallest average size of farms, while Greece holds the lowest position with 4.3 ha.

Greek agriculture is characterized by relatively low levels of mechanization. This also applies more or less to other Southern EU countries. For example, the number of tractors per thousand ha of annual crops is smaller than that of Germany or the Netherlands.

According to Table 1.4, it is interesting to note that all the five Mediterranean countries have 47.3% of the total EU population and the 78.3% of the total EU economically active population in agriculture. As for the total GDP, it seems that Greece shares the lowest levels with Portugal and Spain (albeit at slightly higher levels) compared to the other countries. The same can be said about agricultural GDP per employee, particularly for Greece, with an average of 33 as compared with 100 of the total average of the EU. Greece holds the highest share in the total GDP (11.4%) in comparison with all the other countries. This fact, coupled with the highest shares of economically active population in agriculture (25%), and the highest agricultural exports

(33%), underlines the great importance of this sector for the Greek economy.

On the basis of the above analysis, it can be concluded that a large gap still exists in agriculture between Greece and most of the other EU countries, particularly concerning structure and level of GDP. However, Greece and the other four Mediterranean countries of the EU have similarities rather than differences in almost all the agricultural characteristics referred to above.

Table 1. 4: Population, employment, and agricultural development in EU member-countries

Country	Total population* 1992 (%)	Economically active population in agriculture 1992 (%)	Total GDP 1991 (EU=100)	Agr. GDP per worker 1986 (EU=100)	Share of agr. GDP in total GDP 1990 (%)
GR	2.8	25.0	49	33	11.4
FR	15.5	6.2	115	132	2.8
IT	15.7	11.0	106	66	3.0
PT	2.7	20.0	60	-	4.1
ESP	10.6	16.1	80	-	4.0
BE	2.7	3.0	108	267	1.7
DEN	1.4	6.4	103	169	3.4
GER	21.8	5.2	106	101	1.1
IRE	1.0	13.8	72	98	7.2
LUX	0.1	3.0	131	119	1.7
NE	4.1	4.6	104	263	4.1
UK	15.7	2.5	98	147	1.2
AUS	2.1	-	-	-	-
FIN	1.4	-	-	-	-
SWE	2.4	-	-	-	-
Total EU	100.0	7.2	100	100	3.8

Source: Eurostat * Total population: 367,893,000

Main Issues Stemming from CAP and the WTO

The main problems created in the EU from the implementation of the CAP prior to 1992 have been product surpluses, the high budget, weak structural support and the apparent fragile relationship between agriculture and environment. The 1992 CAP reform, based on the MacSharry plan, focused on the control of product surpluses and budget, farm income support, increasing the competitiveness of agriculture and, more importantly, laying the foundations of a sustainable agricultural development. In regard to this latter point, the policy reform has been completed by environmental policy with the so called «accompanying measures»: these refer particularly to agri-environmental protection, afforestation of farm lands, organic farming, extensive farming systems, stocking densities. So, the last five years sees a new era under way concerning the management and development of the rural sector.

The CAP reform of 1992 has to some extent created favorable conditions for the GATT agreement. Beyond the basic issues of trade liberalization, several parts of the Marrakech Agreements explicitly mention the environment. In view of recent panel decisions, the use of trade barriers for agricultural products based on non-ethical or non-environmentally friendly production techniques will not be allowed by the WTO. The consumers, though, seem increasingly sensitive to concerns of health, environment, and ethics. Therefore, integrated development today must combine the usual goals of innovation, productivity, and quantitative development with fundamental values such as resource conservation and protection, biodiversity, food diversity, and ethics (Mahé, 1997).

A rough estimate of the impact of the evolution of policy in Greek agriculture, over the last five years, since the MacSharry reform, has as follows. Directly, there were rather low rates of development due to product price freezes, while the subsidies and the slight increases in quotas had clearly favorable effects. Indirectly, the policy for the environment and early retirement had rather favorable effects. Sustainable development seems to fit the structure of Greek agriculture quite well, provided that integrated action is taken mainly on reviving traditional techniques of a sustainable character. Infrastructure, systems of marketing and processing activities, should also be improved. Ensuring that people are informed and trained with regard to recent developments will be an important additional consideration.

It is also worth mentioning some important conclusions by Sarris et al (1996): (a) the 1992 CAP reform has introduced the philosophy of direct support to farm incomes, while producers' prices are left to drop gradually to international levels; (b) the WTO will not seriously affect Greek agriculture, which can, on the contrary, utilize some comparative advantages to survive and compete in the European and international market.

A further consideration is to take into account the following main issues of «Agenda 2000», which formulates concrete proposals for reshaping the CAP for the next century (European Commission, 1997):

(a) Consumer requirements from agriculture and food industry in particular will be evolving, due to population growth and rising incomes.

(b) A major aim for the future is the creation of complementary or alternative income and employment opportunities for farmers and their families, on-farm and off-farm; the rural areas are multi-functional, and the farmers should be encouraged to exploit all opportunities for rural entrepreneurs.

(c) A rural policy is becoming very crucial, as the rural areas have increasingly important environmental and recreational functions to fulfil. A prominent role will be given to agri-environmental instruments, by the so called «targeted agri-environmental measures», to support the sustainable development of rural areas and respond to an increasing demand of society for environmental services. On the whole, rural development policy in the EU will aim to integrate market policy, structural policy and environmental policy, in order to establish the overall coherence.

The New Challenges

The high shares of the economically active population in Greek agriculture (22.9%), of agriculture in the GDP (12.5%), and of agricultural exports in the value of total exports (33.0%) illustrate that the economic development of agriculture is important for Greece.

Under the reality of the implementation of the CAP and the GATT agreement, however, it would be difficult to shore up farm incomes indefinitely in the future through price support and income supplements. Emphasis must be gradually placed on a long-term action concerning improvement in agricultural structure, farm enterprise mix, and product quality, in conjunction with a decisive squeezing of production costs. Thus, the main challenges facing Greek agriculture today are related to restrictions created by recent agricultural policies. The various options in agriculture are no longer easy, and the opportunities for expanding farm enterprises are limited. Of course, the challenges concern farmers, farm managers, advisers, specialists, researchers, policymakers, and others related to agriculture to different degrees.

The following challenges constitute new and important parameters to be taken into consideration for planning a re-organization of primary agricultural production in Greece in order to overcome the stagnation in the development process which appears to exist currently.

Structural changes

This is the most important challenge as it refers to the improvement of agriculture in infrastructure, structure of farms and agriculture in general, composition of enterprises, production process and use of innovation to raise quality and squeeze costs of products, the marketing system of each product, and, finally, the human factor.

A re-organization of agricultural production

A new farm enterprise mix is required which should primarily seek to drastically change the ratio of crops versus livestock in total product value which is now approximately 68/32. This is to be realized by replacing surplus cash crops by fodder crops in order to support livestock expansion. In this way, it would be possible to improve the degree of self-sufficiency in livestock products, to limit the value of exchange outflow for importing such products, and to steadily increase the employment and income of the rural population. Of course, expansion of livestock production is meant within the limits imposed by the availability of national natural resources (mainly arable land and pasture land) and their proper and rational use, in order to ensure ecological sustainability.

Product quality improvement and squeezing of production costs

It is not enough to rely on yield levels and possibilities for expansion to face competition. It is also necessary to increase quality of products and reduce costs of production, in order to handle the question of product price more effectively. When supply exceeds demand and where government support schemes operate, surpluses develop, then quality becomes important (Nix, 1987). However, a new approach has to be learned: "What is it the consumer wants - and how much will he or she pay for it?" (Hellard, 1987).

Added-value products

Of particular interest are possible products for processing not merely in industrial factories but also on individual farms with the aim of helping maintain and increase income. This is a very important challenge and has to be encouraged, as it means retention of more and more small farmers and workers in the countryside for the general benefit of rural communities.

Expansion of "organic farming"

It appears that in many countries today the market for "natural" products is expanding rapidly. A differentiation of a price premium for such products together with cost cuttings (fertilizers, spray chemicals, seed dressings, etc.) helps compensate for likely lower yields obtained from organic farms.

Ensuring the future of the countryside

The expansion and development of the traditional livestock enterprises, particularly in less favored and mountain areas, in connection with some other complementary or supplementary activities (like added-value products, organic farming, perhaps tourism, etc.) will lead to a better utilization of natural resources. As a result, a minimum but sufficient human population would remain in such areas, capable of ensuring the protection of the natural environment and hence the future of the countryside.

Conclusions

This chapter has examined the changes to the organization of primary agricultural production in Greece. It emphasizes on its structure and development, particularly before and after accession of the EU, as well as the most important challenges to be faced in the future under the conditions of rapid changes in policy due to the recent CAP reform and the WTO agreement.

On one view, the structure has been hardly affected particularly in the post-entry period due to poor relevant policy measures and funds. However, the economic development continued in the post-entry period but at a much lower rate than in the pre-entry period. This was the result of more or less favorable changes in combination of enterprises, yields, and product prices, as directed by the implementation of the CAP.

In the future it will be difficult to have protection through price support and income supplements. The new challenges facing agriculture now seem to be rapid structural changes, new re-organization of production, improvement in quality and reduction in production costs, creation of added value products, gradual expansion of organic farming, and ensuring the future of the countryside.

References

Demoussis, M. and A. Sarris, 1988. «Greek Experience Under the CAP: Lessons and Outlook,» *European Review of Agricultural Economics*, 15(1):89-107.

European Commission, 1997. «Agenda 2000», Strasbourg.

Eurostat, Various publications (1990-1993).

Georgakopoulos, T. 1988. «The Impact of Accession on Agricultural Incomes in Greece,» *European Review of Agricultural Economics*, 15(1): 79-88.

Hellard, A. 1987. «Over the Gate and Far Away: A Marketing Perspective,» paper presented at the conference on Farming Into the Twenty-first Century, organized by Norsk Hydro Fertilizers, UK.

Mahé, L-P. 1997. «Environment and Quality Standards in the WTO: New Protectionism in Agricultural Trade? A European Perspective,» *European Review of Agricultural Economics*, 24(3/4): 480-503.

Mattas, K. 1990. «EEC's Effects on Greece's Trade,» in van der Noort (ed), *Costs and Benefits of Agricultural Policies and Projects*, Wissenschaftsverlag Kiel KG.

National Statistical Service of Greece, Various publications over the period 1966 -1995, Athens.

Nix, J. 1987. «Appraisal of Current and Future Challenges Facing the Farm Management Specialist,» *Canadian Journal of Agricultural Economics*, 35(4): 675-690.

Sapounas, G. 1994. «Factors Influencing the Declining Contribution of the Agricultural Sector to the Overall Growth of the Greek Economy 1950-1990,» *Spoudai*, 44(3/4): 131-150.

Sarris, A., G. J. Mergos, and P. Sarros, 1996. *The Uruguay Round Agreement for International Trade and its Impact on Greek Agriculture*, Foundation of Economic and Industrial Studies, Special Studies No. 25, Athens.

Zioganas, C. M. 1993. «Structure and Development of Greek Agriculture Before and After Accession to the EC,» *Quarterly Journal of International Agriculture*, 32(2):147-159.

Chapter 2

The Role of Public Infrastructure in the Agricultural Sector

Christos Pantzios and *Panos Fousekis*

Introduction

With the establishment of the World Trade Organization (WTO), as the world appears to be moving towards a freer agricultural trade regime, questions around competitiveness become front line issues in the relevant research and policy agendas. However, the data required for quantitative analysis is not yet available, so that studies addressing these questions will necessarily remain exploratory in nature. Among the key variables influencing the competitiveness of agriculture is productivity (Barkema, Drabestott, and Tweeten, 1990). Productivity is dependent on investment decisions made by the farmer, as well as investment decisions made by the state, including public infrastructure. Competitiveness is often achieved at the expense of environmental sustainability, and public pressure calls for action in this arena too. Environmental sustainability in agriculture is usually approached using qualitative analysis.

The recent WTO agreement on agriculture calls for a reduction in the direct level of domestic support by 20percent on the part of 'developed' member states, leaving exempt only a small percentage of products as well as decoupled income support, investment aid, assistance for land retirement, and payments under environmental programs (Scott, 1996). This implies a loss of competitiveness and an anticipated improvement in environmental conditions. The question is whether other measures not prohibited by the agreement could be employed to boost productivity growth in the agricultural sector. One such measure is public infrastructure, i.e. a public input in production.

The contribution of public infrastructure to the performance of a private firm is typically assessed by the elasticity of private output with respect to public infrastructure. Various studies have attempted to estimate this elasticity with dissimilar results (Aschauer, 1989; Eberts, 1986; Munnell, 1992; McGuire, 1992; Garcia-Mila, McGuire, and Porter, 1996).

The environmental impacts of public infrastructure in the agricultural sector are often dealt with in a qualitative manner, although modern techniques based on cost-benefits analysis are available. Wetland conversion, a typical public infrastructure activity in Greece, has been the focus of extensive research in other countries by environmental economists (e.g., Krutilla and Fisher, 1985; Costanza, faber, and Maxwell, 1989; Hanley and Spash, 1993). It appears that there is a discrepancy between private and social benefits from wetland conversion to agricultural land (Van Vuuren and Roy, 1993). This implies that government decisions to support the private sector that do not incorporate social costs are bound to be inefficient.

This chapter examines the role of public infrastructure as a policy tool in assisting the Greek agricultural sector to become more competitive and environmentally sustainable. Public infrastructure includes mainly land improvements through land drainage and irrigation projects, the construction of some structures, and the use of machinery. The first section, which follows, briefly presents the environmental impacts of public infrastructure policies of the 1970s and 1980s. The next section assesses empirically the contribution of public infrastructure to total factor productivity (TFP) growth in Greek agriculture during the period 1975-1995. Finally the concluding section discusses policy implications in light of an overall assessment of the role of public infrastructure in Greek agriculture.

Public Infrastructure and Its Environmental Impacts

Agricultural policies of the 1960s and 1970s, as well as the five-year national development plan of 1976-1980, emphasized the need for stable incomes through various farm support schemes, correction of structural deficiencies, and higher productivity (KEPE, 1976). These plans and policies dealt with structural deficiencies by promoting increases in the production of export oriented crops, and dealt with productivity via larger farm sizes, mechanization, new seeds, and investments in public infrastructure.

To increase farm size and boost productivity, the five-year national development plan of 1983-1987 proposed measures such as land drainage, land appropriation and distribution, and the founding of agricultural cooperatives (KEPE, 1986). Thus, throughout the 1980s, investments in infrastructure continued to rank high among policymakers' agendas in agriculture.

Unfortunately, public infrastructure has had a catalyst role of a direct and indirect nature in disturbing natural ecosystem functions (Lekakis and Kousis, 1994). Land drainage and multipurpose water projects designed for irrigation have been threatening Greek wetlands. Land conversion projects, from wetlands to cropland, eliminate the potential of wetlands to moderate floods and absorbing toxic substances such as pesticides. In addition, these projects cause unfavorable impacts on wildlife habitat.

Irrigation projects are a prerequisite for the successful planting and cultivation of many modern seeds, which are water and agrichemicals intensive. Irrigated land increased from 13.3 percent of overall agricultural land in 1961 to 36percent in 1993 (Beopoulos and Skuras, 1997). This was the beginning of the intensification process. Corn, sugar beets, and mostly cotton, which are known for their erosiveness and chemical intensity, constitute the main crops irrigated by Greek farmers. Pollution of rivers and lakes by pesticides has been detected by the authorities (Ministry of Environment, 1991). Layers of dead fish and aquatic plants have become a frequent phenomenon in lakes such as *Vistonida* and in rivers such as *Loudias* and *Aliakmon*. Finally, public health problems appear to be serious. However, the magnitude of social costs from environmental damages caused by public infrastructure in Greek agriculture is not known, simply because evaluation studies do not exist.

Public Infrastructure and Productivity

Evaluation of the impact of public infrastructure on agricultural productivity requires deriving and estimating the contribution of this input to total factor productivity (TFP) growth.

Consider a competitive firm with production function,

$$(1)\quad Y(t) = F(X(t), G(t), t)$$

where $Y(t)$ is the level of output obtainable given the flows of n private perfectly variable inputs, $X(t)$, the stock of public infrastructure, $G(t)$, and the state of technology, t. When the partial derivative of F with respect to G, F_G, is strictly positive (negative), public infrastructure is a positive (negative) externality to the firm. When the partial derivative of F with respect to t, F_t, is positive (negative), technical change is progressive (regressive).

The firm selects the levels of the private perfectly variable inputs in order to maximize one-period profits. The firm's problem can be formally stated as

$$(2)\quad \Pi(p, w, G, t) \;=\; Max_X(p\,F(X, G, t) \;-\; w'X)$$

where p is the price of output, and w is a nx1 vector of prices of the perfectly variable inputs. Π is the restricted profit function, which gives the maximum profit attainable given the prices, the stock of public infrastructure and the state of technology.

The first order condition for restricted profit maximization is

$$(3)\quad pF_X \;=\; w.$$

Since the stock of public infrastructure is given to the firm exogenously, G is a parameter in the optimization problem (2). Differentiating both sides of (2) with respect to G and using the first order conditions obtains,

$$(4)\quad \Pi_G \;=\; pF_G.$$

Π_G is the marginal shadow value of the public capital stock, that is, the change in profits due to a change in the stock of public infrastructure. From equation (4), it follows that the marginal shadow value of G is positive (negative) when the marginal physical product of G is positive (negative). Differentiating both sides of (2) with respect to t and using the first order conditions, one obtains

$$(5)\quad \Pi_t \;=\; pF_t.$$

The RHS of equation (5) is the change in profit due to a change in the state of technology. From equation (5) follows that the change in profit is positive (negative) when technical change is progressive (regressive).

The rate of *TFP* growth for the private firm equals the rate of growth in output minus the rate of growth in the aggregate private input. Under the assumption of profit maximization, the rate of growth in the aggregate private input can be calculated as the weighted average of the rates of growth of the individual private inputs, weights being the shares of these inputs in total revenue (Ball, 1985; Chambers,

1989). Formally, the rate of *TFP* growth, denoted here as *RGTFP*, for the firm is,

$$(6) \quad RGTFP(t) = \hat{Y}(t) \; - \; \sum_{i=1}^{n} \frac{w_i(t)X_i(t)}{R(t)} \hat{X}_i$$

where R is revenue and symbol hat stands for logarithmic derivative with respect to time (rate of change).

The rate of change of output, $\hat{Y}$, however, is

$$(7) \quad \hat{Y} \; = \; \sum_{i=1}^{n} \frac{w_i(t)X_t(t)}{R(t)} \hat{X}_i + \frac{F_G(t)G(t)}{Y(t)} \hat{G}(t) + \frac{F_t(t)}{Y(t)}.$$

Substituting (7) into (6) leads to

$$(8) \quad RGTFP \; = \; \frac{F_G(t)G(t)}{Y(t)} \hat{G}(t) \; + \; \frac{F_t(t)}{Y(t)}.$$

Relation (8) suggests that the firm's rate of *TFP* growth can be expressed as the sum of two components. The first component is the primal rate of technical change, $\frac{F_t(t)}{Y(t)}$. The second component represents the effect a change in the stock of public infrastructure on private *TFP* growth. When $\hat{G}$ equals zero (that is, the stock of public capital does not change with time) or when F_G equals zero (that, is a change in G has no effect on output), then the *RGTFP* becomes identical to the primal rate of technical change. When F_t equals zero (that is, there is no technological change), then the change in *TFP* must be attributed to changes in G only.

Substituting (4) and (5) into (8) obtains an alternative expression of the *RGTFP* as

$$(9) \quad RGTFP \; = \; \frac{\Pi_G(t)G(t)}{R(t)} \hat{G}(t) \; + \; \frac{\Pi_t(t)}{R(t)}.$$

The first term on the RHS of (9) captures the impact of public infrastructure on private *TFP* growth while the second term captures the impact of technological change. The former depends on the shadow share of G in revenue. The shadow share of G in revenue is the elasticity of *TFP* with respect to G. We refer to the first term as public infrastructure effect on *TFP* while to the second term as technical change effect.

The Empirical Model and the Data

To estimate the two components of private TFP growth for the agricultural sector in Greece we assume that this sector can be modeled as an aggregate competitive firm that produces one output, Y, using three perfectly variable inputs X_l (labor), X_m (materials), and X_k (private capital), and one public input, G (stock of public infrastructure in agriculture). Y includes crop and livestock production, X_l includes hired and family labor, while X_m includes feed, seed, energy, agrichemicals and other miscellaneous inputs. X_k includes private buildings, structures, and machinery, while G includes land improvements, structures, and mechanical equipment provided by the State to the Greek farmers.[1]

The data sources are as follows. Gross investment series for private capital and public infrastructure are available from the National Accounts of Greece. The perpetual investment model has been used to derive stocks of private capital and public infrastructure employing depreciation rates reported in 'Fixed Capital and Investment in Greek Agriculture' (Agricultural Bank of Greece, 1989, pp. 46-49). The price index for private capital has been derived by dividing investment in current prices by investment in constant (1970=1) prices as suggested by Jorgenson and Yun (1991). The rental price of private capital has been calculated assuming a constant real rate of return to capital of 4 percent.[2] Output and material quantities and the respective price indexes (1970=1) are available from the National Accounts of Greece. The price index for labor (1970=1) is available from the National Statistical Service of Greece. Expenditure on hired labor and the quantity of both hired and family labor (in annual units) are available from the 'Economic Accounts for Agriculture and Forestry', published by *Eurostat*. Expenditure on total agricultural labor has been calculated assuming that the opportunity costs of hired and family labor are identical. The same assumption has been used in earlier studies of the agricultural sector in Greece (e.g., Mergos, 1991).

The profit function is specified as Normalized Quadratic (NQ). The price of private capital has been used for normalization. We denote as p, w_1, w_2, the normalized prices of Y, X_l, and X_m, respectively. Given this notation, the NQ profit function can be written as[3]

$$(10)\quad \Pi(p, w_l, p_m, G, t) = a_o + a_p(p) + a_1(w_1) + a_2(w_2) +$$

$$a_g(G) + \frac{a_{pp}}{2}(p^2) + \frac{a_{11}}{2}(w_1^{\,2}) + \frac{a_{22}}{2}(w_2^{\,2}) + a_{p1}(pw_1) +$$

$$a_{p_2}(pw_2) + a_{12}(w_1w_2) + a_{pg}(pg) + a_{1_g}(w_1G) + a_{2g}(w_2G) +$$

$$a_{pt}(pt) + a_{1t}(w_1t) + a_{2t}(w_2t) + a_{gt}(Gt).$$

The output supply function and the input demand functions for labor and materials are obtained from (10) using Hotelling's Lemma. The demand function for private capital is obtained residually from the relation

$$\Pi = pY - w_1X_l - w_2X_m - X_k$$
$$\Leftrightarrow X_k = pY - \Pi - w_1X_l - w_2X_m.$$

This system of output supply and input demand equations was estimated using the SURE procedure in the TSP program. Asymptotically efficient parameter estimates along with the corresponding t-statistics are reported in Table 2.1. The empirical model appears to fit the data reasonably well. The adjusted coefficients of determination range from 0.7 for the output supply equation to 0.98 for the labor demand equation. The DW-statistics range from 1.42 for the demand for private capital equation to 2.51 for the output supply equation.

The estimated profit function is decreasing in w_l and w_2, increasing in p and convex in w_l, w_2, and p. The derivative of Π with respect to G, Π_G, is positive everywhere (mean 0.262, standard deviation 0.075). [4] A positive shadow value of G suggests that Greek farmers have been benefited from public infrastructure in the sector. The coefficient a_{pg} is positive implying that increases in G lead to higher output. The coefficients a_{1g} and a_{2g} are negative suggesting that public infrastructure is a complement to labor and materials (recall that Hoteling's Lemma gives the negative of the input demand functions). The derivative of the demand for private capital with respect to G is positive (mean 0.207, standard deviation 0.025) suggesting complementarity between the public and the private capital.

Table 2.2 presents the decomposition of private *TFP* growth in Greek agriculture during the period 1975-95 into public infrastructure effects and technical change effects. Table 2.2 also presents the two components of the public capital effects, namely, the shadow share of

G and the rate of growth in public infrastructure in order to illuminate the forces lying behind the evolution of this source of productivity growth over time.

TABLE 2. 1: Estimation results

Coefficient	Estimate	t-Statistic
a_o	4627.5	4.37*
a_p	526053	0.59
a_1	-312750	-1.68**
a_2	-663518	-10.8*
a_g	-7.7	-13.6*
a_{pp}	987.8	1.23
a_{11}	6823.5	1.84*
a_{22}	757.2	2.86*
a_{p1}	-2823.2	-0.76
a_{p2}	539.6	-0.65
a_{12}	-1921.3	-3.04*
a_{pg}	1.39	2.06*
a_{1g}	-0.57	-4.04*
a_{2g}	-0.21	-4.21*
a_{pt}	-256.1	-0.55
a_{1t}	158.9	1.63**
a_{2t}	334.5	10.4*
a_{gt}	0.038	13.7*

	R^2 (Adjusted)	DW-Stat
Output	0.7	2.51
Materials	0.92	1.79
Labor	0.98	1.60
Private Capital	0.95	1.42

* (**) statistically significant at 5 percent (10 percent) or less.

* (**) statistically significant at 5 percent (10 percent) or less.

The public infrastructure effects are positive because both the shadow share and the rate of growth of G are positive everywhere. As mentioned earlier, the shadow share of G is the elasticity of *TFP* with respect to the stock of public capital. The mean value of this elasticity is 0.147 and its standard deviation is 0.034. Our results, therefore, suggest that an increase in G by 1 percent increased the level of *TFP* in

Greek agriculture by almost 15 percent, on average, during the period 1975-95. This number is similar to the output elasticity with respect to

Table 2. 2: Decomposition of TFP growth

	(1)	(2)	(3)=(1)x(2)	(4)	(5)=(3)+(4)
Year	Shadow Share of G	Rate of Growth in G (%)	Public Infra-structure Effects (%)	Rate of Technical Change (%)	Rate of TFP Growth (%)
1975	0.14	3.95	0.54	0.77	1.31
1976	0.15	3.68	0.56	0.79	1.34
1977	0.16	3.09	0.49	0.76	1.12
1978	0.13	2.43	0.31	0.88	1.12
1979	0.15	2.22	0.34	0.81	1.12
1980	0.98	1.41	0.14	0.98	1.11
1981	0.1	1.08	0.11	0.95	1.10
1982	0.13	1.16	0.15	0.93	1.11
1983	0.13	1.77	0.22	0.92	1.11
1984	0.17	1.86	0.31	0.82	1.11
1985	0.14	1.96	0.27	0.87	1.11
1986	0.10	1.02	0.11	0.93	1.10
1987	0.11	0.45	0.05	0.97	1.10
1988	0.13	0.10	0.015	0.95	0.97
1989	0.13	0.28	0.037	0.97	1.01
1990	0.16	0.048	0.008	0.91	0.91
1991	0.17	0.31	0.052	0.88	0.93
1992	0.16	2.06	0.32	0.91	1.23
1993	0.18	3.17	0.58	0.83	1.40
1994	0.22	1.02	0.22	0.77	0.99
1995	0.22	1.37	0.31	0.77	1.07
			Average Values (1975-95)		
	0.147	1.67	0.24	0.87	1.12

G reported in a number of earlier studies (e.g., Eberts, 1986; Munnell, 1990). It is, however, lower, than that obtained by Aschauer (1989) and higher than those presented by Holtz-Eakin (1994) and Garcia-Mila, McGuire, and Porter (1996).

The rate of growth in *G* exhibits a clear downward trend until the late 1980s which has, to a certain extend, been reversed in recent years. It noteworthy that the rate of growth in *G* during the period 1986-90 has been, on average, less than one tenth of that in the period 1975-1980. The reduction in the rate of growth in *G* has resulted in lower public infrastructure effects on productivity. Private *TFP* increased at a rate of 0.4 percent due to public capital effects during the period 1975-80 but only at a rate of 0.037 percent per year during 1986-90. The public infrastructure effects have a mean of 0.00245 (that is, 0.245 percent per year) and a standard deviation of 0.0018.

The technical change effects (represented by the primal rate of technical change) are positive everywhere suggesting that technical change has been progressive during the period 1975-95. The mean value of the technical change effects is 0.0087 (that is 0.87 percent per year) and its standard deviation is 0.00076. A comparison between the technical change effects and the public capital effects reveals that the former are, on average, 3.5 times higher than the later (0.0087 *vs* 0.00245).

The rate of private *TFP* growth, which is the sum of the public infrastructure effects and technical change effects, has a mean of 0.0112 (that is 1.12 percent per year) and a standard deviation of 0.0013. Fousekis and Papakonstantinou (1997), using an adjusted (semi-parametric) Tornquist index report a rate of *TFP* growth of 1.29 percent per annum for the period 1971-93, which is slightly higher than the one obtained in this study. Bureau *et al* (1995) use three non parametric indexes, namely, the Fisher, Hulten, and Malmquist indexes, to calculate *TFP* in Greek agriculture during the period 1973-89. According to the Fisher index the annual rate of *TFP* growth has been 1.75 percent, according to the Hulten index 2.39 percent, while according to the Malmquist 0.52 percent. The differences in the *TFP* growth rates must be attributed to the different behavioral assumptions (profit maximization *vs* cost minimization) as well as to the different methods employed in each study (parametric, non parametric, semi-parametric).[5]

The contribution (in percentage terms) of public capital effects to private *TFP* growth varies considerably with time (mean 0.21, standard deviation 0.136). It ranges a low of 0.8 percent in 1990 to a high of 41.4 percent in 1993. The smallest contribution is observed in the period 1986-90, which corresponds to the lowest rates of growth in *G*. The rate of private *TFP* growth, however, has only been affected slightly by the reduction in public infrastructure effects during 1986-90. This is because the higher rates of technical change during the same period have, to a large extent, offset the reductions in the rate of growth in public capital.

Conclusions

The purpose of this chapter has been to examine the role of public infrastructure as a policy tool in assisting the Greek agricultural sector to become more competitive and environmentally sustainable. The measurement of the contribution of public infrastructure to private efficiency relies on duality results and a decomposition of private *TFP* growth into public infrastructure and technical change effects. The evaluation of the role of public infrastructure in the environmental sustainability of the agricultural sector in Greece is dealt with a qualitative manner drawing upon the findings of earlier works on the topic as well as on official reports.

The main conclusions from implementing the decomposition of private *TFP* growth in Greek agriculture during the period 1975-95 may be outlined as follows: First, public infrastructure has been beneficial to the Greek farmers since higher levels of this public input lead to higher producer profits. Second, the elasticity of private *TFP* with respect to public infrastructure is about 15percent. Third, public infrastructure effects appear to explain about 21 percent of the evolution of private productivity in the agricultural sector of Greece. The remaining 79 percent should be attributed to technological change. Overall, the empirical findings suggest that public infrastructure is productive. In this respect, the empirical results are in agreement with those reported by Aschauer (1989), Eberts (1986), and Munnell (1990) but in contrast with those of Holtz-Eakin (1994), and Garcia-Mila, McGuire, and Porter (1996).

The recent WTO agreements impose a number of restrictions on the policy tools the signatory parties may employ to boost private productivity and, thus, competitiveness of their agricultural sectors. These restrictions do not concern the provision of public infrastructure. Given the large and expanding agricultural trade deficit of Greece and the relatively high elasticity of private *TFP* with respect to public infrastructure, it appears that higher levels of this public input could be justifiable on the grounds of preserving and promoting the agricultural sector's competitiveness. Such a strategy, however, may be incompatible with environmental sustainability. The adverse effects of multipurpose water projects and land conversion projects, which constituted the bulk of public spending on infrastructure in the past, on the functions of natural ecosystems testify to the potential conflicts between the objectives of competitiveness and ecological sustainability. Conflicts of this type may be overcome through a careful selection of appropriate forms of public assistance provided to the farmers. Public investment in agricultural research and in improving the entrepreneurial skills of Greek farmers (including the use of more efficient production and marketing methods) are more likely to achieve the goal of competitiveness and at the same time to ensure environmental sustainability.

Notes

[1] Attempts have been made to include cultivated land in the analysis. They were not successful, however, due to the lack of variability in this production input. The mean value of the cultivated land during this period was 35500 and its standard deviation only 270. Land has been excluded from other empirical production studies in agriculture (e.g., Coyle, 1992; Luh and Stefanou, 1991; Hauver, Yee, and Ball, 1991). The series for the private and the public capital used here involve land improvements which can, somehow, compensate for the exclusion of cultivated land.

[2] To the best of our knowledge, the only studies on the real rate of return to capital are those by Jorgenson and Yun for the U.S manufacturing, and Ball for the U.S. agriculture. Thirtle and Bottomley (1992) suggest a constant real rate of return until more work is done on this elaborate issue. In empirical production analysis, rates of returns from 3 to 6 percent are typically used (e.g., Thirtle and Bottomley (1992); Luh and Stefanou (1991); Vasavada and Chambers, 1986).

[3] The data used for the empirical analysis are aggregate. Consistent aggregation requires that the aggregate profit function is affine in the exogenous input , implying that the second derivative of the profit function with respect to G is zero (Blackorby and Schwarm, 1982; Epstein and Denny, 1983). The aggregation condition has been imposed into the empirical model *a priori*, as in Luh and Stefanou (1991) and Fousekis and Stefanou (1996).

[4] The data used for the empirical analysis are aggregate. Consistent aggregation requires that the aggregate profit function is affine in the exogenous input, implying that the second derivative of the profit function with respect to G is zero (Blackorby and Schworm, 1982; Epstein and Denny, 1983). The aggregation condition has been imposed into the empirical model a priori as in Luh and Stefanou (1991) and Fousekis and Stefanou (1996).

[5] For a thorough discussion about the behavioral assumptions and the methodology underlying the construction of different measures of TFP growth see Chambers (1989) and Bureau, Fare, and Grosskopf (1995).

References

Agricultural Bank of Greece, 1989. «Fixed Capital and Investment in Greek Agriculture,» Athens (in Greek).

Aschauer, D. 1989. «Is Public Infrastructure Productive?» *Journal of Monetary Economics*, 23, 177-200.

Ball, E. 1985. «Output, Input and Productivity Measurement in the U.S. Agriculture,» *American Journal of Agricultural Economics*, 67: 475-86.

Barkema, A., M. Drabenstott, and L. Tweeten, 1990. «The Competitiveness of US Agriculture in the 1990s,» in K. Allen (ed), *Agricultural Policies in a New Decade*, Washington D.C.: Resources for the Future.

Blackorby, C. and W. Schworm, 1982. «Aggregate Investment and Consistent Intertemporal Technologies,» *Review of Economic Studies,* XLIX:595-614.

Beopoulos, N. and D. Skuras (1997). «Agriculture and the Greek Rural Environment,» *Sociologia Ruralis*, 37: 255-69.

Bureau, J., R. Fare, and S. Grosskopf, 1995. «A Comparison of three non parametric measures of Productivity Growth in European and United States Agriculture,» *Journal of Agricultural Economics*, 46: 309-326.

Chambers, R. 1989. *Applied Production Analysis: A Dual Approach*. Cambridge: Cambridge University Press.

Costanza, R., S. C. Faber, and J. Maxwell, 1989. «Valuation and Management of Wetland Ecosystems,» Ecological Economics, 1: 335-361.

Coyle, B. 1992. «Risk Aversion and Price Risk in Duality Models of Production: A Linear Mean-Variance Approach,» *American Journal of Agricultural Economics*, 74: 849-859.

Eberts, R. 1986. «Estimating the Contribution of Urban Public Infrastructure to Regional Growth,» Working Paper No. 8610, Federal Reserve Bank of Cleveland.

Epstein, L. and M. Denny, 1983. «The Multivariate Flexible Accelerator: Its Empirical Restrictions and an Application to U.S. Manufacturing,» *Econometrica,* 51: 647-70.

Eurostat. «Economic Accounts for Agriculture and Forestry,» various issues.

Fousekis P. and A. Papakonstantinou, 1997. «Economic Capacity Utilization and Productivity Growth in Greek Agriculture,» *Journal of Agricultural Economics* , 48: 38-51.

Fousekis P. and S. Stefanou, 1996. «Capacity Utilization under Dynamic Profit Maximization,» *Empirical Economics*, 21: 335-59.

Garcia-Mila, T., T. McGuire, and R. Porter, 1996. «The Effects of Public Capital in State Level Production Function Reconsidered,» *Review of Economics and Statistics,* LXXVII, 177-180.

Hanley, N. and C. Spash, 1993. *Cost-Benefit Analysis and the Environment*, Cheltenham, UK: Edward Elgar.

Hauver J., Yee J., and E. Ball, 1991. «Capacity Utilization and Measurement of Agricultural Productivity,» Technical Bulletin 1798, Economic Research Service , USDA.

Holtz-Eakin, D. 1994. «Public Sector Capital and the Productivity Puzzle,»

Review of Economics and Statistics, 76: 12-21.

Jorgenson D. and K. Yun, 1991. *Tax Reform and the Cost of Capital*, Oxford: Clarenton Press

KEPE (Center of Planning and Economic Research), 1976. *Crop Production*, a report for the 5-year development plan of Greece, 1976-80, Athens (in Greek).

KEPE, 1986. *The Primary Production Sector in Greece*, a report for the 5-year development plan of Greece 1983-87, Athens (in Greek).

Krutilla, J. V. and A. C. Fisher, 1985. *The Economics of Natural Environments*, Washington D.C.: Resources for the Future.

Lekakis, J. N. and M. Kousis, 1994. « Agriculture and the Environment in Greece,» *Progress in Rural Policy and Planning*, 4: 177-207.

Luh, Y. and S. Stefanou, 1991. «Productivity Growth in U.S. Agriculture Under Dynamic Adjustment,» *American Journal of Agricultural Economics*, 73: 116-25.

McGuire, T. 1992. Highways and Macroeconomic Growth: Phase two. Federal Highway Administration (March).

Mergos, G. 1991. *Output Supply and Input Demand in Greek Agriculture: A Multi-Output Profit Function Approach*, Athens: Center of Planning and Economic Research.

Ministry of Environment, 1991. «National Report of Greece,» prepared for the UN Conference on Sustainable Development in Rio, Athens.

Munnell, A. 1990. «How does Public Infrastructure Affect Regional Economic Performance?» *New England Economic Review*, Federal Reserve Bank of Boston, Sept./Oct., 11-32.

Munnell, A. 1992. «Policy Watch: Infrastructure Investment and Economic Growth,» *Journal of Economic Perspectives*, 6, 189-98.

Scott, J. 1996. «Tragic Triumph: Agricultural Trade, the Common Agricultural Policy, and the Uruguay Round,» in N. Emiliou, and D. O'Keefee (eds). *The European Union and the World Trade Law After the GATT Uruguay Round*, N. York: John Wiley and Sons.

Thirtle, C. and P. Bottomley, 1992. «Total Factor Productivity in the U.K. Agriculture, 1967-90,» *Journal of Agricultural Economics*, 43: 381-399.

Van Vuuren, W. and P. Roy, 1993. «Private and Social Returns from Wetland Preservation versus Those from Wetland Conversion to Agriculture,» *Ecological Economics*, 8: 289-305.

Vasavada, U. and R. Chambers, 1986. «Investment Demand in the U.S. Agriculture,» *American Journal of Agricultural Economics*, 68: 960-970.

Chapter 3

Crop Production Under Risk-Uncertainty Conditions

George Vaphiades

Introduction

The analysis of risk-uncertainty in the farm's production process lies in the front line of international economic literature. From a practical point of view, also, the issue of risk is of major importance to the Greek agricultural sector since many crops have reached their production limits. Consequently, agricultural policy makers need to know the farmers' behavior with respect to risk-uncertainty, which plays a decisive role in income variability, in view of market globalization and increased competition.

Agricultural risk, whether it refers to systematic factors or non systematic ones, is an important dimension of the production process because it focuses on the instability or variability of: quantity produced which is affected by climatic changes, diseases and other natural causes, the producer price for the crop, and/or, the prices of the inputs. Since farms are subjected to risk factors and have to operate under different circumstances of uncertainty, any changes in these factors cause corresponding changes to producers choices and, possibly, to consumers decisions which in turn could influence general market equilibrium.

The theoretical analysis of risk and uncertainty as well as any relevant empirical investigations are based on utility theory (Newbery and Stiglitz, 1981). Producers try to maximize the expected value of profit (EP) or income (EY) when they are risk-neutral. On the other hand, when they are risk averters or risk lovers they seek to maximize the expected utility of income EU(Y). In both cases, apart from the mathematical formulation of the utility function, one has to estimate crop income within a multi-period context.

So far as the quantitative analysis of farms' behavior under risk is concerned there are two kinds of approach in the international literature. According to the first one the functional form of the income utility function is traced out by experimenting on the responses of a number of farms to various hypothetical or actual risky alternatives. The main shortcoming of this direct approach is that farmers are not properly motivated to make the right choice in view of the fact that many alternatives are hypothetical and are not related to any cost evaluation. Furthermore, even if farms are faced with real alternatives, the ensuing choices are static in the sense that their behavior under risk cannot be empirically investigated over a number of years.

The second category puts the emphasis on a more indirect approach whereby the parameters of a proposed model are estimated on the basis of previous farms' decisions with respect to risky alternatives for a certain period. In this way, their behavior under risk can be deduced by evaluating the explanatory power of the model in question. The main problem with the second approach is that empirical results are totally dependent upon the a priori specification of the mathematical model.

Due to the subjective nature of risk measurement in agriculture, evaluation of risk attitudes has been a rather difficult task. As a result there have been only a limited number of relevant models. One of them was suggested by Moscardi and de Janvry (1977) who assumed that farmers minimize the probability of income reduction below a minimum level at a certain period (safety-first objective), that is they maximize the quantity $EY\text{-}K\sigma_Y$, where EY is the expected value of income, K is a parameter and σ_Y the standard deviation of income. The authors used a Cobb-Douglas production function, observations on fertilizer inputs, and other socio-economic information from a sample of forty five farms in order to investigate risk attitudes.

Schluter and Mount (1976) estimated average yields and producer prices for a period of six years in the Surat District in India. Then, they used observations from thirty three farms to estimate the mean income and its mean absolute deviation for the sample period. By formulating a linear programming problem, the authors tried to trace out the trade-off between the above two in determining the optimal plan for the representative farm (risk-efficient farm plan) and for different levels of uncertainty.

The mean-variance model, which is used in the present chapter, has been widely employed in the literature due to its basic assumption that the distribution of income is normal and, therefore, it can be fully described by two parameters[1]: the mean and the variance. The normality assumption implies that utility functions have a certain functional form, that is $U(Y) = -Ke^{-AY}$, where Y is income and K, A are parameters (Newbery and Stiglitz, 1981). In this case the coefficient of absolute risk aversion is constant and the slope of the corresponding indifference curves depends only on income variance. Thus, if the mean of the income distribution remains constant, risk is solely dependent upon the variance of the distribution.

Buccola and Subael (1984) suggested another criterion in order to rank probability distributions related to alternatives characterized by low risk avertion. Instead of applying the mean-variance criterion, they proposed the mean-Gini coefficient criterion which bares no consequences to the functional form of the utility function. More specifically, assuming that one has to choose between two distributions F(x) and G(x), corresponding to two different alternatives, there will be a preference for the former if $\mu F \geq \mu G$ and $\mu_F - \Gamma_F \geq \mu_G - \Gamma_G$, where μ_F, μ_G are the means and Γ_F, Γ_G are the Gini coefficients of the two distributions. An additional assumption for this decision criterion is that one of the above expressions should be a strict inequality.

Hertzler (1991) pointed out that there are certain state variables, such as the farm's wealth, which are dynamic in nature and their change can be described by a stochastic process. The farmer's decisions or choices at the beginning of a time period [τ-t] entail a certain risk since they can induce random changes to the farm's accumulated net profits and, therefore, change the farm's wealth. In this context, wealth is treated as a stochastic variable which evolves in time and its fluctuations are described by the following stochastic difference equation: $W\tau - Wt = f(t, Wt, Ct, \xi t)\,[\tau - t]$, where C is a vector of controlled variables, ξ is the vector of the stochastic process and t, τ are the present and future times respectively. Assuming that the difference equation can be approximated by a differential one in continuous time, the stochastic process has the Markovian property (white noise) and that events (i.e., change in producer prices) take place at a relatively fast rate dW can be expressed by an Ito differential equation. Wealth accumulation is stochastic in nature due to income risk resulting from the farm's choices. Thus, a risk averter farmer can maximize an objective function of expected utility EU(W) under the constraint of the differential equation dW (Ito stochastic control). According to Hertzler, the main advantage of this method lies with the fact that it is equally robust to the mean-variance criterion without the restrictive assumptions of the latter, that is the normality of the random variable's distribution or the specific functional form of the respective utility function.

Finally, in a recent discussion paper, Baltas and Korka (1994) investigated producers' choices with respect to land uses at the Amvrakikos Bay Area. They adopt a utility function with a wealth variable defined as the sum of initial wealth plus the random return of every farming activity.

Although risk in agriculture is related to various price stabilization schemes, future markets, buffer stocks management, and other means of dealing with price uncertainty, issues which frequently appear in the international literature, the present chapter focuses on the empirical investigation of producers' attitude to risk in the Greek agricultural sector within a multi-period context. It is assumed that the above attitude implies that producers are more interested in crop income variability than producer price instability. There is no doubt that in the last decade institutional prices under CAP have greatly reduced the latter. The chapter follows the indirect methodological approach to risk-uncertainty, according to which income, consumption and wealth are treated as random variables following a certain distribution determined by the functional form of the corresponding utility function. The latter reflects the agent's behavior in the face of risk. In essence, the agent (farmer) is shown to choose any particular distribution on the basis of the criteria proposed by the model. The estimated parameters of the chosen distribution are described in terms of these criteria and they are, sometimes, used in maximization problems.

Applying the Mean-Variance Model

The Mean-Variance model is applied to yearly income from a sample of fifteen selected crops over the period 1979-1990. To make sure that the mean-variance analysis in not misleading, the normality of the ensuing income distributions is subjected to testing. In this case, it is implied that the producers are risk averters and risk is uniquely related to crop income variance. A marginal case of the above model is risk neutrality. There is no doubt that producers attitude to risk differs from region (area) to region, but due to lack of data on production cost, it was not possible to make the analysis region specific. Therefore, the results of the present empirical investigation refer to the whole agricultural sector. This aggregation shows the general tendency, but conceals the existing differences from region to region. In essence, the application of the mean-variance model, in a multi-period context, entails the estimation of two parameters, the mean and the variance, for each crop of the selected sample for the period 1979-1990.

The second stage of the empirical investigation involves the estimation, using factor analysis, of a composite parameter representing the trend of certain random variables which characterize the dynamic evolution of the selected crops during the period in question, such as acreage, yield, cost of production, output, price etc. Moreover, an attempt is made to classify the evolutionary production pattern of the fifteen crops by applying three criteria. More specifically, for each crop the estimated composite growth rate for the area allocated to its production, its share in the overall acreage and its yield are used to classify the fifteen crops into increasing, stationary and declining ones. This classification is additionally investigated by applying cluster analysis on the three criteria.

Since for each crop there is an estimated trend parameter representing the change in the above mentioned n random variables, there are n (15 x 1) trend vectors which constitute a matrix (15 x n). At this stage of the empirical investigation, the multivariate analysis of correlation and partial correlation are applied in order to test the correlation between the trend parameters of the n or n-p random variables. By doing so, an attempt is made to answer two main questions: firstly, to see how far the correlation between the production pattern of the crops in question, based on the three criteria, and the corresponding income mean and variance support the utility assumptions related to farmers' risk aversion or risk neutrality (farmers response to risk). Secondly, to trace out the contributing factor(s) to income risk ; that is the factor(s) that explain the variance of the yearly income from the fifteen selected crops. One has to point out that the estimated values of the correlation coefficients are not important per se, what is of interest is their statistical level of significance (α being not greater than 10%) and their ranking order. For this reason, the corresponding non parametric coefficients of correlation (Kendall, Spearman) are estimated.

The Data

The data employed are annual observations covering the period 1979-1990 for a sample of fifteen representative crops the gross output of which amounted to about 47%-49% of the overall crop gross output. For each crop the data refers to changes in the following variables: cultivated area (acreage), acreage structure for the fifteen crops (share changes), yield, cost of production, nominal producer prices including income subsidies, and gross output per 1,000m^2. The first three variables are the criteria for crop classification into increasing, stable and declining ones. On the other hand, cost of production, producer prices and yield are the main sources of income risk.

The above period was chosen for two main reasons. Firstly, it was the period just before and after Greece's accession to the EEC (now EU) when major adjustments took place in the agricultural sector. Secondly, the only available data on the cost of production referred to the years 1978, 1980, 1981, 1987 and 1989 (see Zioga, Delis, and Schortsanitis, 1992). For the remaining years of the time series the cost values were proxied by projecting the annual growth rate from the relevant subperiods. The data on the other variables was taken from the Agricultural Economic Statistics computed by the research and planning division of the Agricultural Bank of Greece. For certain products, such as tobacco, raisins, peaches, oranges, edible olives and winegrapes, the estimated yields refer to nonmarginal plots of land. Peaches and potatoes are divided into three subproducts or qualities (early peaches, spring potatoes etc.) characterized by different yields and prices. In order to aggregate the relevant data on these crops, the not too unrealistic assumption was made that the average cost of production is the same for every quality. So far as yield and producer prices are concerned, the computed aggregate values are the weighted average of the individual values with each subproduct's annual quantity being the corresponding weight.

In order to draw as much information as possible from a set of data and estimate a reliable trend for each of the above listed variables, three methods were used. First an annual growth rate was calculated by using the compound interest formula, $Y_t = Y_0 (1+i)^t$, or in the case of the period in question the applied formula becomes $Y_{1990} = Y_{1979} (1+i)^{11}$, where Y_{1990} is the observation on the variable in 1990, Y_{1979} the initial value and i the annual rate of growth. The basic shortcoming of this method is that the estimation of i is based on the two end values which might not be representative of the period 1979-1990. Particularly when the evolution of certain variables, such as crop production cost, yield or prices, is examined one might get the wrong picture or pattern. Another method which was employed to estimate the growth rate was the application of OLS to an exponential model with time being the exogenous variable. Finally, the third method to trace out any long term changes can be described as follows: the period in question was divided into two subperiods, that is 1979-84 and 1985-90, which have the same number of observations. This division does not necessarily mean that the two subperiods have any distinct characteristics. Then, the average value of every variable was calculated for each subperiod and the ratio $(\text{mean}_{1985\text{-}90} / \text{mean}_{1979\text{-}84})\%$ was computed. Thus, the percentage change of the two subperiod averages could indicate some more permanent changes in the listed crop variables.

The next step is to estimate a composite trend parameter by applying factor analysis to the three computed growth rates. Thus, for each variable there is a trend parameter vector (15 x 1) corresponding to the fifteen crops. In addition, at the final stage, the trend vectors, as well as the vectors of mean crop income and income variance, are employed for the estimation of the correlation and partial correlation coefficients.

It should be noted that crop data which expresses average values for all nonmarginal producers in the country may conceal existing deviations due to different farm profiles (size, land quality, area etc.). In the present chapter, a not too restrictive assumption was made that average values satisfactorily describe the profile of the average nonmarginal producer.

The Empirical Results

Generally speaking, attitudes to risk are described in terms of utility theory which, irrespective of the methodological approach, implies that the choices of agents are subjective in a microeconomic sense and that they may differ from region to region. In this respect the empirical examination of their aggregate behavior is a limitation of the present study. However, any investigation of producers' attitude to risk on the aggregate may turn out to be very useful in the sense that it reveals the behavior of the average viable farm (not operating on the margin) in a multi-period context on the satisfaction of the following methodological assumptions:

i) A rationally behaving farm should be focusing on the variance of crop income and not on the variance of producer nominal prices.

ii) Crop income is a random variable, normally distributed over the period 1979-90, a hypothesis which was not rejected by the two-sided Kolmogorov-Smirnov test for the fifteen selected crops, as seen in Table 3.1.

iii) For each of the random variables characterizing each crop of the sample, there is a composite trend parameter which describes the evolution of the random variable over the period 1979-90.

iv) The above parameter has been estimated by applying factor analysis to three values, one estimated by using OLS to the sample data, the second calculated on the basis of the familiar compound interest rate formula and the third by computing the percentage change between two sub-periods of the sample.

The multivariate analysis of correlation and partial correlation of the various composite trend parameters have given the following results:

Nominal producers prices for the fifteen crops are not correlated to the corresponding quantities. The coefficient was not statistically significant and was found to be equal to - 0.15. This lack of correlation

TABLE 3. 1: Two-sided K-S Test of normality for crop income

Basic crops	Estimated Kolmogorov D plus and D minus values	Estimated Kolmogorov DN Maximum Value	Level of Significan ce
Durum Wheat	0.180 and 0.240	0.240	0.490
Maize	0.157 and 0.077	0.157	1.000
Cotton	0.159 and 0.182	0.182	0.999
Tobacco (Basmas)	0.117 and 0.142	0.142	1.000
Oranges	0.138 and 0.142	0.142	1.000
Edible Olives	0.187 and 0.163	0.187	0.999
Sugar canes	0.230 and 0.248	0.248	0.451
Common barley	0.207 and 0.195	0.207	0.999
Potatoes	0.174 and 0.231	0.231	0.542
Hay (a-a)	0.171 and 0.215	0.215	0.999
Currants	0.126 and 0.275	0.275	0.322
Peaches	0.146 and 0.078	0.146	1.000
Wine grapes	0.132 and 0.294	0.294	0.247
Soft wheat	0.185 and 0.232	0.232	0.536
Brewery barley	0.178 and 0.188	0.188	0.999

Note: The critical value of the test for α=0.05 and n=12 is 0.375 and since for all crops DN<0.375 all the empirical distributions appear to have a good normality fit.

implies that the average Greek farmer does not try to maximize mean crop income: that is the hypothesis of risk neutrality is not substantiated empirically. This result could have various alternative or complementary explanations:

• During the first years of the period in question and just after Greece's accession to the EEC in 1981, there had been a significant increase in nominal prices, whereas after 1987, the prices became more or less stable with small positive increases. On the other hand, the corresponding quantities produced, either due to climatic conditions or due to the imposition of quotas just after 1987, did not follow the same pattern.

• With certain crops in Greece (Apostolou and Varelas, 1987) the quantities produced are influenced to a large extent by nominal prices of competitive products or by the general price index (relative prices).

• It is very likely that when it comes to producers' choices they take into consideration income subsidies, nominal prices with a time lag and/or withdrawal prices. One should also note that Greek farmers do not easily change their production pattern. Oranges are a characteristic example of farmers' inflexible behavior. The nominal price of oranges declined considerably over the period in question while the growth rate of the quantity produced was low but positive.

The standard deviation of crop income appears to have a statistically significant and positive correlation only with the cost of production. Therefore, the latter appears to be the only contributing factor to income risk.

There is positive correlation between the cultivated area of each crop (and its corresponding share in the overall acreage structure) and its yield. It seems that an increase in the cultivated area increases the overall productivity resulting in economies of scale.

Gross output per 1,000 m^2 in also positively correlated to income mean and variance. Therefore, an increase of gross output per 1,000 m^2 leads to an increase in the mean of the distribution of crop income for the sample crops. However, the cost of production increases at the same time raising the variance of the distribution, which means that any change in expected utility is inconclusive according to the mean-variance model. Assuming that crop income follows the normal distribution, any increase in the mean would raise expected utility (less risk) if the standard deviation remained the same. It is worth noting that the estimated partial correlation coefficients show that there is a significant positive relation between gross output per 1,000 m^2 and income mean when income variance remains stable. This case, however, implies that yield increases with the cost of production held constant, that is there is an increase in total factor productivity.

There is no correlation between income crop variability, $\sigma_y / \bar{Y}$ and any of the three variables (acreage, acreage structure and yield) that could be possibly used as a criterion in order to classify the fifteen crops into declining, stable and increasing ones. Income crop variability appears to have a negative correlation coefficient (-0.21) with gross output per 1,000 m^2 but it is not considered as statistically significant. The multivariate analysis of correlation proved to be inadequate to provide any empirical hints with respect to the above crop classification which is of great importance to policy makers. For this reason cluster analysis was applied to the data on the aforementioned variables by using three different classification algorithms (Jobson, 1992). The results indicated that 20% of the sample crops had been increasing over the period 1979-90, namely durum wheat, maize and cotton. Oranges and potatoes (13.3%) had been more or less stable and the remaining ones (66.7%) were on the declining side.

Although the absolute values of the estimated income variation coefficients are of no importance, a comparative assessment of these values leads to some interesting results. The average value of the income variation coefficients for the fifteen crops of the sample is equal to 154, as in Table 3.2.

TABLE 3. 2: Crop income (1979-1990)

Basic crops Ranked by the Growth Rate of Cultivated Area	Annual Growth Rate of Cultivated Area	Coefficient of Variation $\sigma_{n-1}/\overline{X} \bullet 100$	Mean Income $\overline{X}$ (Drs/ 1000 m^2)	Std Deviation σ_{n-1}
Durum Wheat	+9.3%	227	-958	2,174
Maize	+9.2%	112	1,703	1,912
Cotton	+7.0%	71	5,245	3,742
Tobacco(Basmas)	+0.8%	355	2,542	9,024
Oranges	+0.7%	108	14,157	15,233
Edible Olives	+0.2%	131	9,835	12,905
Sugar canes	+0.15%	430	-1,019	4,386
Common Barley	-0.2%	83	-2,330	1,941
Potatoes	-1.4%	150	-6,595	9,926
Hay (a-a)	-1.6%	112	-1,916	2,146
Currants	-2.0%	146	-17,616	25,772
Peaches	-2.2%	99	21,040	20,782
Wine grapes	-2.8%	72	-4,656	3,347
Soft wheat	-5.7%	118	-1,808	2,127
Brewery barley	-8.6%	97	-1,910	1,856

Amongst the increasing crops, cotton has the lowest coefficient (71), durum wheat the third highest one (227) - although its mean income appears to be negative - and maize has a coefficient substantially below the average (112). Oranges and potatoes, the two stable crops, are also characterized by below the average values, 108 and 150 respectively. Again, potatoes appear to have a negative mean income over the period in question. With the sole exception of durum wheat, the four non-declining crops have low coefficients of income variation, something that could signify that Greek farmers are risk averters. Alternatively, the estimated mean income for durum wheat and potatoes was negative, whereas there appear to be declining crops, such as edible olives and peaches, with positive mean income and low coefficients of variation (131 and 99 respectively). It seems that cluster analysis, as well as the ranking of crops on the basis of their coefficient of income variation, confirm the multivariate analysis of correlation main result, that is the average Greek farmer does not seek to maximize the expected utility of crop income.

Conclusions

The empirical investigation of Greek farmers' attitude to risk-uncertainty within a multi-period context seems to reject the hypothesis that the average farmer is either risk-neutral with respect to crop income risk, or that his choices focus on the maximization of the corresponding expected utility. This conclusion does not necessarily mean that expected utility does not play any role in the average farmer's crop choices and production decisions. In this case further research is required towards the identification of the functional form of utility by introducing additional variables, such as total household expenditure and/or wealth, particularly when agricultural households receive income from sources other than farming.

The problem of empirically investigating farmers' attitudes to risk in a multi-period context is far more complicated than the static case. If farmers can transfer income from one year to the next and if they rationally choose a long term consumption pattern, then their behavior under risk becomes more complex. In a multi-period context, risk attitudes depend not only on the shape of the utility function, but also on initial wealth, the discounting rate, the degree of independence of successive risks and future income opportunities (Newbery and Stiglitz, 1981).

Due to the recent GATT agreements, with the aim of further reducing trade barriers in world trade, it is expected that competition will increase in the international markets of agricultural products and inputs. Consequently, the revision and gradual elimination of the CAP and the related institutional prices, even if direct income subsidies persist, will reinforce, in the long run, the competitive conditions for Greek agricultural products and, in essence, it will increase crop income risk.

Furthermore, it is likely that in the context of a sustainable growth of Greek agriculture any future environmental regulations will increase the cost of production. Thus, the statistically significant and positive correlation between the latter and crop income risk, as the present chapter has indicated, means that any environmental regulations, in addition to trade liberalization, will enhance the tendency towards higher risk in crop production.

Finally, in the years to come, risk management will be of vital importance to crop production choices. As far as Greek agriculture is concerned, policy makers should try to inform and convince the average Greek farmer about the importance of income changes and the associated risk when it comes to crop choices in the long run. At the same time, further research is required in order to empirically investigate farmers choices between risky alternatives.

References

Agricultural Economic Statistics, Research and Planning Division, Agricultural Bank of Greece (annual edition).

Apostolou N. and E. Varelas, 1987. *Supply Functions of Selected Crops,* Agricultural Economic Studies No 25, Agricultural Bank of Greece Publication,

Baltas N. and U. Korka, 1994. «Modelling Farmers' Land Use Decisions,» *Discussion Papers* No 59, Department of Economics, Athens University of Economics and Business, September.

Buccola S. T. and A. Subael, 1984. «Mean - Gini Analysis, Stochastic Efficiency and Weak Risk Aversion,» *The Australian Journal of Agricultural Economics*, 28(2-3): 77-86.

Flury B. and H. Riedwyl, 1988..*Multivariate Statistics-A Practical Approach,* New York: Chapman and Hall.

Hanson S. and W. G. Ladd, 1991. «Robustness of the Mean-Variance Model with Truncated Probability Distributions,» *American Journal of Agricultural Economics*, 73(2): 436-445.

Hertzler G. 1991. «Dynamic Decisions Under Risk: Applications of Ito Stochastic Control in Agriculture,» *American Journal of Agricultural Economics*, 73(4): 1126-1137.

Jobson, J. D. 1992. *Applied Multivariate Data Analysis*, New York: Springer Verlag.

Moscardi E. and A. de Janvry, 1977. «Attitudes Toward Risk Among Peasants: An Econometric Approach,» *American Journal of Agricultural Economics*, 59(4): 710-716.

Newbery D. and J. Stiglitz, 1981. *Theory of Commodity Price Stabilization. A Study in the Economics of Risk*, New York: Oxford University Press,

Schluter M. G. G. and T. D. Mount, 1976. «Some Management Objectives of Risk Aversion in the Choice of Cropping Patterns, Surat District, India,» *Journal of Development Studies*, 12(3): 246-261.

Zioga N., D. Delis, and K. Schortsanitis, 1992. *Cost of Production of Agri cultural Products and Efficiency in Greek Agriculture (1969-1989)*, Agricultural Economic Studies No 43, ABG Publications.

Chapter 4

Farm Household Adjustment to the Increasingly Competitive Conditions in European Agriculture

Sophia Efstratoglou

Introduction

Since the middle of the eighties, market surplus of agricultural products and heavy budgetary spending at European level initiated substantial changes in the support for agricultural products and resulted in various reforms of the CAP. Basic objectives of these reforms have been to reduce market imbalances, restrict budgetary spending and increase the competitiveness of EU agriculture both in domestic and external markets (Tracy, 1996).

Under the pressure of the Uruguay Round, the MacSharry reform of the CAP in 1992 initiated substantial reductions in prices for arable crops and beef so as to bring them closer to international price levels and achieve market balances. The GATT agreement signed in 1994 initiated a 20% reduction in domestic support of agriculture over a six year period, a reduction of 36% in budget spending on export subsidies and a 21% cut in the quantity of subsidized exports (European Commission, 1997a). The prospect of WTO Trade talks in 1999 will further reduce domestic support and put extra pressure on European agricultural markets and farm incomes (European Commission, 1995).

Agenda 2000 announced in 1997 stressed the new objectives that should guide future CAP developments, and specifically the need to "improve competitiveness of EU agriculture on both domestic and external markets" (Ahner, 1997), and proposed new reforms in the cereals, beef, dairy and Mediterranean products regime, aiming at more balanced markets through reduction in support prices.

Within this framework, the Greek farming sector, which still remains a vital part of the national economy, has been faced with increasingly competitive conditions and with declining support.

Greek agriculture contributes 14% to the formation of GDP (1995) and employs 20% of the total labor force, compared to the EU averages which are respectively 3% and 5.3% (European Commission, 1997b). The sector is characterized by weaknesses such as small farm size, low productivity, aging labor force with lower managerial skills and a generally slow modernization process (see also Zioganas, this volume). Almost half of the farms with low productivity are located in mountainous and hilly areas and are low productivity farms.

Exodus from farming has been accelerating during the last decade following Greece's accession to the Union (1981-1991). Agricultural labor force declined by 32% and the number of farms decreased only by 13% (the corresponding declining rates during 1971-1981 were 22% and 5%).

What remains also a significant problem is the low presence of 'business-professional farms in agriculture', farms which are characterized by larger farm sizes and higher productivity. In Greek agriculture, these farms account only for about 25-30% of the total number of farms (Efstratoglou, 1994).

The purpose of this chapter is to identify farm households adjustment patterns and survival strategies in the context of increasingly competitive conditions in the European agriculture.

Farm Household Adjustment Strategies

The identification of the adjustment patterns of the Greek farm households to their changing context has been carried out in two study areas, *Korinthia* and *Fthiotis*.[1] A sample of 625 farm households (318 in Korinthia and 307 in Fthiotis) was formed by two level stratification: farm size and location (level - hilly - mountainous areas). To trace and interpret changes and behavioral responses, observations and findings from three basic research tools (and stages) were utilized: the Baseline Survey of 625 farm households in 1987, the Final Survey in 1991 and three Panel Surveys in between. The Baseline and the Final Survey provided the necessary quantitative and qualitative information on the structural characteristics of the farms surveyed and the changes that occurred in them during 1981-1991, while the Panel Surveys on 130 farm households out of the total, explored through the farm life-history approach further qualitative aspects in farm households adjustments processes. Changes observed and recorded concerned the following key aspects: (a) changes in resource use, investment undertaking and business adjustments at farm level, (b) changes in family labor allocation among farm and non farm activities, (c) response to policy measures, awareness, knowledge and reasons behind use and non-use of policies, and, (d) succession and the problem of non-continuity in farming.

Finally, a cluster analysis of the main features and changes observed at farm household level during 1981-1991 was carried out (Efstratoglou, 1992). From this analysis, three types of farm households were identified with regards to adjustment patterns and strategies followed. These were: the *Professionalization* Strategy, the *Disengagement* Strategy, and the *Stable Reproduction* Strategy. The three different patterns of behavior also provide a dynamic typology of farms which is extremely useful for policy considerations. More analytically:

The Professionalization strategy refers to farm households that are engaging more resources in farming and have a positive attitude towards it. Most farmers through this engagement process are adjusting to market changes, are intensifying and specializing in farming and are responding to modernization policy incentives (Professionalization typology).

The Disengagement strategy characterizes farm households that have been disengaging resources from farming for reasons such as age, lack of succession, small farm sizes, and orientation towards non-farm activities. In these farm households, the importance of farming declines over time (Disengagement typology).

The Stable Reproduction farm strategy concerns farm households that tend to reproduce a stable situation in farming with minor changes in resource use and with a lack of response to changing market conditions, to policy measures and generally to structural change (Stable Reproduction typology).

Furthermore, within each basic strategy, sub-categories were identified. Specifically, within the Professionalization strategy two subgroups of farm households were observed: (a) those with a business professionalization strategy towards farming ('conventional productivist professionalization') and (b) those that were engaging resources in farming but at lower level of professionalism ('low enterpreneurship'). Also, within the stable reproduction strategy, three different groups of farms were identified: (a) those whose stable reproduction was due to a satisfactory level of accumulation and farm income ('post-professionalization stability'), (b) those that farming was a supplementary activity with dominant pluriactivity (multiple job holding); and, (c) those whose farming is at marginal level (survival strategy) just before starting the disengagement process from it (Efstratoglou, 1990a).

The percentage distribution of farm households in the two study areas by their dynamic typology (strategy) is presented in Table 4.1. The main elements, characteristics and changes observed under these strategies are further analyzed below:

Table 4. 1: Distribution of farm households by strategy adopted

Strategy	Korinthia (%)	Fthiotis (%)
1. Professionalization	25	19
-Conventional professionalization	(12)	(9)
-Professionalization with low entrepreneurship	(13)	(10)
2. Disengagement	23	29
3. Stable Reproduction	52	53
-farming at satisfactory level of accumulation	(8)	(17)
-farming as a supplementary activity	(17)	(19)
-farming as a survival strategy	(27)	(17)
Total	100	100
Number of surveyed households	305	303

Source: Final Survey 1991, European Research Programme on "Rural Change in Europe: Farm structures and pluriactivity 1987-1991".

The Professionalization Strategy

Within this strategy, the main characteristic is further commitment of resources in farming. Two types of behavioral response were observed: the 'conventional professionalization' strategy and the 'engaging in farming with low entrepreneurship' strategy.

Conventional professionalization

Under this sub-category, farm households which are depending on farming are engaging more resources in it and have a very positive attitude towards it. Since 1981, most of them have increased their farm size through the purchase or lease of land and some have committed extra family or hired labor resources to it. Capital investment undertaken in farming by these farm households has allowed further specialization and intensification of production and has increased the productive capacity of the farm e.g. irrigation systems, green houses, machinery, buildings, livestock, etc.

These farmers have been price responsive and market-oriented. Crop restructuring was highly observed among these farmers as a response to changing market conditions and farm prices. Farmers under that typology had a very good knowledge of policy measures and were responsive to modernization incentives (797/85, or 355/77, or National law 1262/81). They could be characterized as "the dynamic farmers and business professionals" in the Greek agriculture and some of them have developed leadership initiatives at local level (co-operatives, local government etc.). Some of them have been experimenting with new crops and were early adopters of new technology.

Most of them were monoactive (only in farming), but whenever pluriactivity was observed in these households, most of the time it was centered around farming activity which tended to be vertically integrated agricultural activities - that is, retail marketing of their farm products or work on other farms with their machinery on a contractual base. In both cases such behavior implies a more efficient use of the labor and capital available on the farm. In our survey analysis, 12% of the farm households in Korinthia and 9% in Fthiotis were found to be under this sub-category (Table 4.1).

Professionalization with low entrepreneurship

In this sub-category, farm households also depended on a high degree on farming and had a commercial attitude towards it. What really characterized them is that they have been engaging resources to farming; labor, land and capital. They leased or purchased land to the degree that local conditions allowed it, and allocated more family labor on the farm. They were expanding farm activity and investing, but their investment levels were much lower when compared with the previous sub-category of farms. The main difference from between the "business professionals" and this group is their lower level of knowledge of policy measures and of perception of technological and market changes taking place in their environment. They were price responsive and were restructuring crops as a result of price changes (or subsidies changes), but no ability was observed to perceive new market prospects or respond to structural measures. Most of them have had self-financed farm investment or have partly used credit. Reasons reported for not having used investment subsidies were lack of knowledge of policy measures, risk averting attitude, uncertainty, bureaucratic procedures, etc. Despite their positive attitude to farming, and the fact that they were engaging resources, it is evident that their entrepreneurial and managerial capacities are at lower levels compared to the 'business professionals.' All of them had a farming background, a sense of continuity and tradition in farming and tried to create favorable conditions for succession. Younger farmers (<55 of age) believed that at least one of the younger family members will would succeed them, while the older ones already had a successor in farming. The percentages of these sub-categories were 13% in Korinthia and 10% in Fthiotis (Table 4.1).

'Non productivist' professionalization

Under the Professionalization typology, no farm households were observed to be adapting to a 'non-productivist' professionalization behavior" where farming is related to new environmental concerns such as extensification schemes, set aside, low-input production, and organic farming. This is attributed to the fact that these practices have dimensions not easily perceived by farmers, unless systematic work and promotion of them is undertaken by local policy actors. Many of these measures (set-aside, extensification, organic farming, reafforestation etc.) have only been initiated since 1989 within the relevant CAP regulations, and it was evident that implementation of these measures had not been promoted enough by policy makers. Moreover, most of these measures do not appeal as very relevant in many Greek rural areas, where low farm incomes and poor structures prevail and priorities on increases in productivity and cost reduction still remain vital. Recently though, extensified practices in farming, mainly organic farming, seem to be promoted within the CAP reform.

The Disengagement Strategy

The main feature of these farms is that they are withdrawing resources from farming and disengaging from it. Most of these farms are owned by elderly farm households (of two aging persons) with no succession. However, under this typology, a small number of younger farm households were also found. All of them gradually remove land from farming and reduce family labor input, but for different reasons. Most of the elder farm households are in their late life cycle and their disengagement strategy is related to their age and lack of succession. Most of the time, younger family members have moved out of farming and out of the area, too. The small farm size did not allow the reproduction of the farm household and its survival and 'pushed' younger family members to migrate when local economies could not retain them.

With regard to the younger farm households disengagement, this is due again to the small farm size that has 'pushed' them towards an off-farm orientation for supplementary employment and incomes.

Older farmers under the disengagement typology were not responsive to policy measures, while younger farmers were not entitled to them whenever farming was not their main activity.[2]

Disengagement should be differentiated from retirement. In most of the cases of the farm households surveyed, farm operator's retirement did not imply a disengagement process when succession existed. Only in cases where there was not a successor was the disengagement process associated with retirement and exit from farming. The proportion of farm households under the disengagement typology was found to be 23% in Korinthia and 29% in Fthiotis (Table 4.1).

The Stable Reproduction Strategy

The basic feature of this behavioral adjustment pattern is that the farm households tend to reproduce a stable situation in farming. This typology covers 52% of the farm households in Korinthia and 53% in Fthiotis (Table 4.1). In other words, half of the farm households in both areas remain static and do not respond to structural change and changing market conditions. There are only changes occurring at the farm level such as minor labor allocation or investment undertaking, mainly for replacing worn out capital. These minor changes aim at maintaining the stable reproduction of the farm activity. Another common element of this behavioral pattern is that in many cases, surplus from farming has been transferred out of agriculture and has been invested in off-farm activities, in non-farm assets etc.

Although the elements of this typology are common, the causes of it present differences among the farm households. Three main subcategories were identified within this typology: (i) farming being at satisfactory level of accumulation and farm income (post professional stability); (ii) farming being a supplementary activity for the farm household (dominant pluriactivity); and, (iii) farming being a survival strategy.

Post professionalization stability

These are farm households with a high degree of dependence on farming and most of them are in the late family cycle (farm operators over 55 years). Their behavioral pattern of response has been static during the last decade (since 1981) as a result of the accumulation of resources (land and capital) achieved in previous stages of the family cycle through an engaging or business professionalization adjustment pattern.

Their farm sizes and incomes are above average and the marginal changes occurring at the farm level are mainly crop restructuring in response to price and subsidies changes. Their knowledge of policy measures varies from average to poor with no interest in the use of structural measures. They are monoactive (only in farming) farm-households and younger family members are already established outside of farming. This has been achieved partly through transfers of agricultural surplus to investment in human resources. This group of farms does not respond to modern technology or to modernization incentives. This type of behavioral pattern is observed in 8% of the farm households in Korinthia and in 17% in Fthiotis (Table4.1).

Farming being a supplementary activity with dominant pluriactivity

Farm operators under this sub-category have their main gainful activity outside farming. In most cases, farm incomes from farming vary between 25-40% of the total family income. Farms are of small or medium size and were inherited.

Farm operators have a farm family background, but they left farming at a younger age and were employed in the secondary or tertiary sector. Some of them exploited job opportunities offered by the local labor markets, or if such opportunities were lacking, they migrated. Upon inheriting the farm, they started farming as a supplementary activity for the following reasons: either for keeping the farm land as an asset or for attachment to farming and continuity of tradition or as a hobby. The asset approach to farming seems to be frequently observed, particularly in rural areas where farm land is highly priced because of its potential alternative uses.

Most of these farmers have a poor knowledge of policy measures. Some of them are not interested in them and none are entitled to them, since farming is not their main occupation. They are only consuming market subsidies, and they have a low market orientation and price responsiveness. They are also pluriactive farm households with both types of pluriactivity (farmers/spouse and other family members), while most of the farm operators are very often self-employed in services. It is a rather stable adjustment pattern over time and is observed in 17% of farm households in Korinthia and in 19% in Fthiotis.

Farming being a survival strategy

This typology is observed in 27% of the farm households of Korinthia and 17% in Fthiotis (Table 4.1) and consists mainly of farm households of small family size and small farm size (marginal farms). They have a high or medium degree of dependence on farming, and they remain in farming in a static situation, by reproducing survival. Marginal changes on the farm concern crop restructuring as a result of price or subsidy changes. The level of knowledge of policy measures is inadequate and although they are entitled to the use of measures, they do not consume them. Reasons recorded for non use of policy measures were age, lack of successor and small farm size. Younger family members have moved out of farming and exploited job-opportunities either in their local labor markets or migrated. The possible evolution pattern of adjustment of this type of farm households over time is either disengagement, if there will not be a successor, or farming as a supplementary activity if there will be a successor.

Factors Influencing Farm Households Adjustment Patterns

Factors influencing farm household adjustment strategies and structural change at the farm household level were the "core" of the Arkleton research program. The hypothesis states that two types of factors are influencing farm household behavior, the internal and external factors (Arkleton Research, 1992):

(a) Internal factors refer to the farm and family characteristics such as farm size(s), labor-capital availability, skills, farmer's age, family cycle, and prospect of succession. Non-favorable structures at the farm level e.g. small farm sizes, low farm income are expected to 'push' farm households to off-farm employment and to supplementary sources of income, while older age, late family cycle, lack of succession contribute to a stable reproduction or disengagement strategy.

(b) External or "contextual" factors form the environment in which farm families operate and face opportunities and constrains (agricultural markets and policies, local labor markets, local economic and social organizations). Favorable non-farm labor market structures are expected to 'pull' farm households into non-farm employment and disengage resources from farming.

It has been argued (Arkleton Research , 1990; Efstratoglou, 1990b) that whenever favorable agricultural structures prevail and higher farm incomes and return to resources can be achieved, farm households tend to engage more resources and develop a professionalization strategy in farming. On the other hand, if agricultural structures are weak (non-favorable) farm households tend to disengage resources from farming and try to exploit off-farm labor opportunities, developing in this way a disengaging strategy or as a stable reproduction strategy combining it with pluriactivity.

Based on the above, professionalization patterns are expected to be observed at higher levels in areas with better farm structures compared to areas with non-favorable farm structures. Conversely, higher disengagement is expected in areas with poor farm structures such as small farms, poor land, aged farm operators, etc. (strong 'push' factors).

Empirical evidence from the Greek study areas is consistent with the above expectations. Professionalization was higher (25%) in Korinthia compared to 18% in Fthiotis, as seen in Tables 4.2 and 4.3. The greater percentage of the surveyed farm households in this typology were also in non-Less Favored Areas (LFAs) (74%). Conversely, disengagement was higher in Fthiotis (29%), compared to 23% in Korinthia. More than half of these households in Fthiotis were in the LFA (65%).

The higher professionalisation rates in Korinthia are consistent with Korinthia's contextual structures. Korinthia is an area with more favorable agricultural and non agricultural structures compared to Fthiotis, reflected by the higher farm business sizes (average farm size in Korinthia is 11 ESU[3] -European Size Units- compared to 6.5 ESU in Fthiotis), and in the higher returns per land and labor units because of intensive farming systems (Efstratoglou, Papageorgiou, and Pliaka 1987a, 1987b). On the other hand, less favorable farm structures in Fthiotis seem to 'push' more households into disengagement strategies or stable reproduction ones with the presence of pluriactivity (Table 4.3).

Stable reproduction patterns were high in both areas (52% and 53% respectively). These findings are at about the same levels with adjustment patterns observed for Group A rural areas in Arkleton's comparative analysis (Arkleton Research, 1992). The highest rates of stable reproduction (65%) among all types of areas were recorded for Group A which consisted mainly of Southern European study areas with higher dependence on farming. At the same time, engaging patterns in Group A areas were at about 18% and disengaging patterns at 16%.

With regard to "internal factors" at farm household level, it has been assumed that age, farm size, family cycle and succession are key variables affecting the adjustment strategies of farm households. From Tables 4.2 and 4.3, one can observe the differences in these key variables among farm households under different adjustment strategies.

Table 4. 2: Main characteristics of farm households surveyed (FH) by type of strategy in Korinthia.

Characteristics	Prof.	Dis.	St. R.	Total
1. Number of FH	76	70	159	305
Percent of FH	25	23	52	100
2. Percent of FH in Less Favored Areas	26	48	30	33
3. Degree of dependence on farming (%):				
High (% agr. income >75)	30	11	20	24
Medium (% agr. Income = 25-75)	66	30	52	51
Low (% agr. income <25)	4	51	28	26
4. Average dependence on farming (%)	64	31	48	48
5. Average farm size in ha.	14.4	2.4	5.5	7.0
6. % distr. of farms by class in ESU*:				
Small farms (<4 ESU)	5	73	33	35
Medium farms (= 4-16 ESU)	42	27	57	46
Large farms (> 16 ESU)	53	-	10	18
7. % of farms by farm operator age:				
Young farmers (< 55 years)	59	7	52	43
Older farmers (> 55 years)	41	93	48	57
(of which without successors)	(58)	(30)	(40)	(42)
8. % Distribution of FH by OGA**:				
only farming	60	53	29	34
farmer/spouse OGA	33	32	65	57
other family member OGA	7	15	7	8
9. Average family size (adults)	4.11	3.75	3.68	3.92
10. Patterns of farm production:				
(% distribution of UAA*** in)				
Arable	47	37	43	-
Citrus and fruits	11	8	13	-
Olive trees	29	41	35	-
Vineyards	6	4	6	-
Other	7	10	2	-
Sheep/goats (number of heads)	80	3	8	-
11. Degree of depend. on social transfers				
no dependence (0%)	49	15	44	38
low (<10%)	41	20	31	31
medium(10-40%)	9	23	15	15
high (>40%)	1	42	10	15

*ESU= European Size Unit, 1 ESU=1200 ECUs
**OGA= Other gainful activities (pluriactivity)
***UAA=Utilized Agricultural Area

Source: Final Survey, 1991, Research Programme "Rural change in Europe: Farm structures and pluriactivity 1987-1991".

Table 4. 3: Main characteristics of farm households surveyed (FH) by type of strategy in Fthiotis.

Characteristics	Prof.	Dis.	St. R.	Total
1. Number of FH	56	87	160	303
Percent of FH	18	29	53	100
2. Percent of FH in Less Favored Areas	36	56	53	51
3. Degree of dependence on farming (%):				
High (% agr. income >75)	54	6	23	24
Medium (% agr. Income = 25-75)	46	26	41	38
Low (% agr. income <25)	-	68	36	38
4. Average dependence on farming (%)	75	21	44	44
5. Average farm size in ha.	20.0	1.8	8.0	8.5
6. % distr. of farms by class in ESU*:				
Small farms (<4 ESU)	7	98	49	55
Medium farms (= 4-16 ESU)	57	2	44	34
Large farms (> 16 ESU)	36	-	7	10
7. % of farms by farm operator age:				
Young farmers (< 55 years)	55	16	46	39
Older farmers (> 55 years)	44	84	54	61
(of which without successors)	(63)	(27)	(43)	(43)
8. % Distribution of FH by OGA**:				
only farming	56	45	24	32
farmer/spouse OGA	28	45	67	59
other family member OGA	6	10	10	9
9. Average family size (adults)	4.2	2.8	3.9	3.6
10. Patterns of farm production:				
(% distribution of UAA*** in)				
Arable	92	66	80	-
Citrus and fruits	-	-	-	-
Olive trees	6	30	16	-
Vineyards	-	-	-	-
Other	2	4	4	-
Sheep/goats (number of heads)	24	3	-	-
11.Degree of depend. on social .transfers				
no dependence (0%)	57	26	46	43
low (<10%)	34	11	32	27
medium(10-40%)	9	22	15	15
high (>40%)	-	40	6	15

*ESU= European Size Unit, 1 ESU=1200 ECUs
**OGA= Other gainful activities (pluriactivity)
***UAA=Utilized Agricultural Area.

Source: Final Survey, 1991, Research Programme "Rural change in Europe: Farm structures and pluriactivity 1987-1991".

Under the professionalization typology, farm households have a large productive base (large and medium size farms account for over 90% of the total number of farms under this typology) and depend significantly on farming (64% and 75% average dependence on farm income in Korinthia and Fthiotis). They are mostly young farmers (<55 years old), and whenever the farm operator is over 55, there is the prospect of succession. Their dependence on social transfers is very low and, although pluriactivity is observed, it is at much lower levels compared to the other two typologies and to the average levels of pluriactivity observed in the area. Labor resource availability is reflected in the greater average family sizes under the professionalization strategy in both areas.

Farm households under the disengaging typology have very small farms. Almost all farms in Fthiotis and 75% of the farms in Korinthia under this typology have a business farm size less than 4 ESUs. Their dependence on farming is very low. In more than half of the farm households, farm income contributes less than 25% to the family income while a significant percentage of the farm households under this typology (40%-42%) depend on social transfers (social transfers account for more than 40% of the family income).

In almost all farm households, under the disengagement typology,the farm operator is over 55 years old and without the prospect of succession while their average family size is the lowest observed (2.6-2.8 family members). Because of the disengaging attitude toward farming, family members are oriented to non-farm employment, and hence, pluriactivity rates were recorded at higher levels compared to professionalizing farm households.

Stable reproducers have a lower dependence on farming compared to professionalizers, and the incidence of off-farm work is at the highest rates observed. Farmers are younger under this typology compared to disengagers, and they have low dependence on social transfers. They tend to have small and medium size farms, but their farms are larger compared to disengagers. Significant percentages of this typology are recorded in LFAs.

Conclusions

Within the increasingly competitive conditions in European agriculture, Greek agriculture has accelerated its transformation process (population exodus from farming and decline in the number of farms). However, the modernization processes seems to be moving very slowly. Constraints in human resources skills, organizational aspects of production, low response to modernization incentives, and low rates of capital formation contribute to the sector's low productivity.

The prospects for Greek agriculture within the changing market environment depend to a great extent on the farmers' capacity to adjust to more competitive conditions. The analysis of the strategies adopted by the surveyed farm households has shown that only a small percentage of farms perceive market changes and respond to modernization incentives. About one fourth of the farms have a professionalization strategy that allows them to intensify production, reduce costs, and adjust to more competitive markets. These are the "business farmers" (or business professionals) of younger age with managerial skills and relatively larger farms.

On the opposite side, the aging problem in Greek agriculture (almost 54% of the labor force is over 45 years old) and the prevalence of small farms result in stable reproduction strategies characterized by lack of response to changing markets, to modernization policies and capital formation. More than half of the farms, for the various reasons explained, have shown low adjustment to changes. Considering the intensification of restrictive price policies (CAP reform in 1992, Agenda 2000) and the further trade liberalization (WTO talks prospects), it is expected that a large part of these farms will be marginalized. Under these pressures, they will have to either exit farming or restructure their resource uses to achieve higher efficiency.

The rather high rates of disengagement strategies observed in Greek agriculture (23-26%) are explained by the aging problem and the small scale of farming (small farms) that do not allow the survival of the farm and of the farm household in the long run. Hence, development strategies for Greek agriculture formulated by policy makers should take into consideration the differences in farm household adjustment strategies and needs as they appear within the changing market environment.

Notes

[1] The European Research Programme "Rural change in Europe: Farm structures and pluriactivity 1987-1991" was carried out by ten member states research teams, coordinated by Arkleton Research during 1987-1992. The scientific responsible for the Greek team was the author of this chapter which draws heavily from the work developed in the project.

[2] Modernization incentives measures and direct subsidies to farmers such as compensatory allowances in less favored areas (LFAs) are granted only to those farmers for whom farming is their main activity (more than 50% of the family income is earned from farming).

[3]The European Size Unit (ESU) is a unit of measurement of the economic size of the agricultural holding. A farm has an economic size of 1 ESU if its total standard gross margin (SGM) is ECU 1200. The standard gross margin for each enterprise corresponds to the average value, over a three-year period and in a given period, of production minus certain variable costs (Decision 85/377/EEC) (European Commission, 1997b).

References

Ahner D. 1997 . «The Future of the CAP Prospects, Problems, Challenges and first Orientations in Agenda 2000,» European Commission, Directorate-General VI Agriculture, Internal Document PACE/EN/(0002.doc).

Arkleton Research, 1990. «Emerging Responses of Farm Households to Structural Change in European Agriculture,» Working Paper, Arkleton Trust (Research), Estone, Oxford, UK (ATR/90/11).

Arkleton Research, 1992. «Farm Household Adjustment in Western Europe 1987-1991,» Final Report on the research programme "Rural change in Europe: Farm structures and pluriactivity, 1987-1991", December 1992.

Efstratoglou, S. 1990a. «Structural Change and Dynamic Household Typologies - An Integrated Analysis of the Adjustment Patterns adopted by Farm Households in Korinthia and Fthiotis,» Second Report to the EEC, The Arkleton Research 1990, p.p. 173-190.

Efstratoglou, S. 1990b. «Pluriactivity in different Socio-economic Contexts: a test of the push-pull hypothesis in Greek farming,» *Journal of Rural Studies*, 6(4): 407-413.

Efstratoglou, S. 1992. «Structural Changes and Developments at the farm household level: the case of Korinthia and Fthiotis (Greece),» Integrated Panel Analysis Report, Research Programme on "Rural change in Europe: Farm Structures and Pluriactivity, 1987-1991".

Efstratoglou, S. 1994. «Structural Development and Prospects of Greek agriculture,» *Express*, Special Issue on Agriculture, October 1994 (in Greek).

Efstratoglou, S., K. Papageorgiou, and A. Pliaka, 1987a. «Context Analysis: Part I, Developments in Agricultural Structures in Korinthia and Fthiotis, Greece,» Report, Research Programme "Rural change in Europe: Farm Structures and Pluriactivity, 1987-1991".

Efstratoglou, S., K. Papageorgiou, and A. Pliaka, 1987b. «Context Analysis: Part II, Developments in Non-Agricultural Structures in Korinthia and

Fthiotis, Greece,» Report, Research Programme "Rural change in Europe: Farm Structures and Pluriactivity, 1987-1991".

European Commission, 1995. *GATT and the European Agriculture*, CAP Working Notes, Special Issue, Brussels.

European Commission, 1997a. *The Common Agricultural Policy: An Overview*, Directorate-General VI for Agriculture, March.

European Commission, 1997b. The Agricultural Situation in the Community, 1996 Report, Brussels.

Tracy, M. 1996. *Agricultural Policy In the European Union and other Market Economies*, Agricultural Policy Studies, Belgium.

Zioganas, C. M. (this volume), «The Organization of Primary Agricultural Production in Greece: Evolution and Challenges for Re-Orientation.»

Chapter 5

Environmental Sustainability and Price Support under Production Uncertainty

Giannis Karagiannis

Introduction

Steadily rising farm prices have resulted in tremendous output growth. This has been associated with highly intensive agricultural activities, which in turn progressively required the use of marginal land (agricultural extensification) and the extensive use of various intermediate inputs (agricultural intensification), especially fertilizers and pesticides. Agricultural extensification brings marginal and highly erodible land in production causing sediment and chemical runoff into surface water. It also affects wildlife habitat and diversity of plants and animals through conversion of forest, wetlands, and other natural features to agricultural uses. Agricultural intensification, on the other hand, tends to increase the use of fertilizers and pesticides, and the concentration of livestock. It may also result in degradation of on-farm resources from salinization, chemical pollution and soil erosion.

Both agricultural extensification and intensification create concerns of environmental as well as health problems in agricultural areas. Given though that most of the output growth in European agriculture has been achieved through yield improvement (Murphy, Furtan and Schmitz, 1993), the latter seems more important in terms of environmental concerns. It has extensively been argued that the higher producers' prices provided incentives to farmers to produce more and in order to do so they increased, among others, the use of fertilizer and pesticides (given that are normal inputs), which are considered to be the main pollution sources in agriculture (deWit, 1988; Abler and Shortle, 1992; Brouwer and vanBerkum, 1997; Lewandrowski, Tobey, and Cook, 1997). This is the so called «intensification hypothesis».

The «intensification hypothesis» has important regional aspects in the EU, as southern countries have followed a model of agricultural development which is different and less intensive than that adopted by northern countries (Beopoulos and Skuras, 1997). During the last three decades, southern European countries have achieved a lower fertilizer/land ratio, but have reached faster annual growth rates (see Table 5.1). The average annual growth rate of fertilizer/land ratio is greater than 1% for all southern countries, Finland and the U.K. Despite this, all southern countries, except France, still have a relatively lower fertilizer/land ratio compared to northern European countries.

TABLE 5. 1: Fertilizer/Land Ratio and its Average Annual Growth Rate for EU Countries, 1961-1993

Country	Fertilizer per Hectare	Average Annual Growth Rate (%)
Austria	0.20	0.28
Belgium-Luxembourg	0.47	0.30
Denmark	0.19	0.96
Finland	0.13	1.34
France	0.16	2.37
Germany	0.36	-0.11
Greece	0.06	4.11
Italy	0.06	4.00
Netherlands	0.63	0.59
Norway	0.21	0.97
Portugal	0.05	1.94
Spain	0.04	3.15
Sweden	0.13	0.27
U.K.	0.20	2.06

Source: FAO

On the other hand, production uncertainty may be a more important source of risk for southern EU countries, as northern countries' agriculture is more livestock production intensive. It is well known that exogenous sources of production uncertainty, such as weather conditions (temperature, rainfall and drought), are more important in crop than in livestock production. It may be argued that crop/livestock and fertilizer/land ratios are positively related to each other, since a faster growth of crop production is more likely associated with higher yields and increased fertilizer use per unit of land.

The objective of this chapter is to test the validity of the «intensification hypothesis,» which asserts that farm sector price supports leads to greater use of fertilizers and pesticides per unit of land (Lewandrowski, Tobey, and Cook, 1997), under conditions of risk and production uncertainty. Production uncertainty is involved in all agricultural activities and it is mainly attributed to exogenous factors as well as to inputs used. It is, however, much more significant in crop than in livestock production. Production uncertainty is considered to be the sole source of risk, as the established price support system in the EU has eliminated all output price uncertainty.

Under these circumstances, the risk-input relations along with their technical relationship (substitutes or complements) is of major concern in analyzing the environmental impact of agricultural policies. The former is mainly associated with profit variability, which is of equal importance with the expected profit for risk-averse farmers. As is extensively discussed in the next section, fertilizer is more likely to increase profit variability whereas the opposite is true for pesticides.

Conceptual Framework

In the preceding analysis, it is assumed that inputs used by a perfectly competitive firm may cause some environmental problems, as well as health damages. It is also assumed that firm's production is stochastic, which may be due to various reasons; that the level of actual output cannot be known with certainty at the beginning of the production period and that the firm has no means to face production uncertainty, e.g. crop insurance.

On the other hand, the government may implement policies, which may reduce or increase pollution caused by firms. In particular, it has been argued that agricultural policies and environmental damages are linked to each other (Just and Antle, 1990; Hrubovcack, LeBlank, and Miranowski, 1990; Leathers and Quiggin, 1991; Karagiannis, 1997b). It is not however as clear, as it may initially be thought, how to link theoretically and empirically agricultural policies, farm practices and environmental problems in rural areas.

Following Just and Antle (1990), the major problem is that environmental impacts associated with various field level operations do not aggregate to the larger geographical units that are relevant to policy issues. Even though the relationship between agricultural production and environmental damages is site specific, determined mainly by farm practices, agricultural policies are issued for larger geographical regions, which in the case of the EU consists of 15 countries. Nevertheless, Just and Antle (1990) have shown that agricultural intensification can directly be linked to environmental damages when the quantity of land in production, and land environmental characteristics are held constant. These assumptions are adopted in the preceding analysis.

Consider a risk-averse perfectly competitive firm facing production uncertainty. Suppose that the firm makes its production decision before uncertainty is resolved and that it has no production flexibility to adjust its output *ex post*. Even though production is uncertain, the cost of production is assumed to be known with certainty as it is measured in terms of planned rather than actual output. The objective of the representative risk-averse perfectly competitive firm is to maximize the expected utility of profit for any given policy implemented.

Production uncertainty is assumed to have the following form (Just and Pope, 1978): $y=f(x)+h(x)\varepsilon$, where y is the planned output, x is a vector of variable inputs, ε is a random variable with $E(\varepsilon)=0$ and $f_x>0$, $f_{xx}<o$ and $h_x>(<)0$ for every risk-increasing (risk-reducing) input. According to Pope and Kramer (1979), an input is said to be marginally risk-reducing (risk-increasing) if, under risk aversion, the expected value of its marginal product is less (greater) than the marginal factor cost. Consequently, a risk-averse farmer uses less (more) of a marginally risk-increasing (risk-reducing) input under production uncertainty than otherwise.

If $h(x)$=constant, the above model is reduced to the case of additive production uncertainty. Additive production uncertainty corresponds to cases where output variability is independent of the level of planned output. This may be the case for uncertainty arising from weather or other natural phenomena affecting agricultural production.

Alternatively, if $f(x)=h(x)$, the case of multiplicative production uncertainty arises, i.e. $y=f(x)(1+\varepsilon)$. In the case of multiplicative production uncertainty, the variance of output is positively related to the level of planned output. As an example, consider the case of pest disease: the crop loss depends on the level of planned output.

The Just-Pope's (1978) model also allows the simultaneous existence of both risk-increasing and risk-reducing inputs in order to take into account differences in product variability caused by inputs like labor and fertilizer on the one hand and like capital, machinery and pesticides on the other. For example, pesticide use increases yields when a pest infestation occurs, but does not affect yield if no infestation occurs. By eliminating the lower tail of the yield distribution, pesticide use will increase mean yield and reduce the variance of yield. Similarly, irrigation increases (or keeps constant) yield when a drought occurs during the crop session, but it does not affect yield when rainfall is adequate and timely.

On the other hand, the use of fertilizer increases the probability of very high yields when rainfall is adequate and timely but also increases the probability of low yields when rainfall is inadequate and chemical burning occurs. Thus, increasing fertilizer use, increases both the mean yield and yield variability. Similarly, hired labor may increase average product when attention and supervision is effective, but it may decrease output otherwise.

An interrelationship exists between the form of production uncertainty and the inputs' risk characteristics. This occurs through the variance-level-effect. According to Magnusson (1969), the variance-level-effect is zero of the expected value and the variance of the random variable is constant. This means that the frequency distribution does not include the decision variable (input level). The forms of additive and multiplicative production uncertainty are included in this case. Moreover, for these two forms of production uncertainty, only the inclusion of risk-increasing is allowed. On the contrary, in the case of random input services there is a non-zero variance-level-effect as the distribution of the random variable depends on the level of decision variables (input use). In this case, the existence of both risk-reducing and risk-increasing inputs is allowed. The general model of Just and Pope (1978), which is used in the proceeding analysis, is such an one.

Theoretical Results

The objective of the representative firm is to maximize the expected utility of profit for any predetermined level of production subsidy, s:

$$\max_{x} Eu(\pi) = Eu\{(1+s)p[f(x)+h(x)\varepsilon] - w'x\} \tag{1}$$

where $u(\bullet)$ is a von Neuman-Morgenstern utility function with $u'(\bullet)>0$ and $u''(\bullet)<0$ for risk averse producers, and p and w are the certain output and input prices, respectively. The first-order conditions to the above optimization for the two-input case are:

$$Eu'(\bullet)\{(1+s)p[f_1(x)+h_1(x)\varepsilon] - w_1\} = 0 \tag{2a}$$

$$Eu'(\bullet)\{(1+s)p[f_2(x)+h_2(x)\varepsilon] - w_2\} = 0 \tag{2b}$$

The second order conditions require the corresponding Hessian matrix to be negative semi-definite, which is satisfied under risk aversion and a quasi-concave production function. The typical elements of the Hessian matrix are:

$$D_{11} = (1+s)pEu'(\bullet)(f_{11} + h_{11}\varepsilon) + Eu''(\bullet)\pi_1^2$$
$$D_{22} = (1+s)pEu'(\bullet)(f_{11} + h_{22}\varepsilon) + Eu''(\bullet)\pi_2^2$$
$$D_{12} = D_{21} = (1+s)pEu'(\bullet)(f_{12} + h_{12}\varepsilon) + Eu''(\bullet)\pi_1\pi_2$$

To analyze the effect from an increase in production subsidy on input use, differentiate (2a) and (2b) with respect to s and solve the resulting system of equations to obtain:

$$\frac{\partial x_1}{\partial s} = -\left(\frac{p}{|D|}\right)(D_{22} - \delta D_{12})\left\{fEu''(\bullet)\pi_1 + h_1 Eu'(\bullet)\varepsilon + \left(\frac{h}{h_1}\right)Eu''(\bullet)\pi_1 h_1 \varepsilon\right\} - \\ -\left(\frac{p}{|D|}\right)(D_{22}f_1 - D_{12}f_2)Eu'(\bullet) \qquad \text{(3a)}$$

$$\frac{\partial x_2}{\partial s} = -\left(\frac{p}{|D|}\right)(\delta D_{11} - D_{12})\left\{fEu''(\bullet)\pi_1 + h_1 Eu'(\bullet)\varepsilon + \left(\frac{h}{h_1}\right)Eu''(\bullet)\pi_1 h_1 \varepsilon\right\} - \\ -\left(\frac{p}{|D|}\right)(D_{11}f_2 - D_{12}f_1)Eu'(\bullet) \qquad \text{(3b)}$$

where $\pi_i = p(1+s)[f_i(x) + h_i(x)\varepsilon] - w_i$.

PROPOSITION 1: *Assuming cost minimization, stochastic complementarity and IPRRA, a production subsidy results in an increase of a risk-reducing input.*

PROOF: The sum of the first two terms in the bracket of (3a) and (3b) is positive for risk-reducing inputs (Karagiannis and Gray, 1996; Lemma 2, p. 459). The third term in the bracket is also positive for risk-reducing inputs (Pope and Kramer, 1979; Lemma 3, p. 493). $D_{22} - \delta D_{12} < 0$ and $\delta D_{11} - D_{12} < 0$ under stochastic complementarity and different risk-input relations. Under cost minimization, $D_{22}f_1 - D_{12}f_2$

reduces to $p(f_1 Eu'(\bullet)F_{22} - f_2 Eu'(\bullet)F_{12})$ and $D_{11}f_2 - D_{12}f_1$ to $p(f_2 Eu'(\bullet)F_{11} - f_1 Eu'(\bullet)F_{12})$, respectively and both are negative as long as F_{12} >0 (stochastic complementarity).

QED

The behavior of the partial relative risk aversion coefficient provides information about the behavior of a proportional change in the risk premium resulting from a given proportional change in risk (Menezes and Hanson, 1970). A sufficient condition for partial relative risk aversion to be increasing in profit is that relative risk aversion is non-decreasing and absolute risk aversion is non-decreasing in profit, with at least one of them being non-constant (Menezes and Hanson, 1970; Zeckhauser and Keeler, 1970). Increasing partial relative risk aversion requires farmers to have a utility function in which marginal utility diminishes at a rapidly decreasing rate and a proportional increase in risk results in a more than proportional increase in aversion to risk.

There is empirical evidence that farmers' risk attitudes are characterized by increasing partial relative risk aversion. Love and Buccola (1991) and Saha, Shumway and Talpaz (1994), for example, have found increasing partial relative risk aversion for Linn Country and Kansas wheat farmers, respectively.

In the presence of the variance-level-effect, a sufficient condition for cost minimizing behavior of a competitive firm under production uncertainty is either risk neutrality, CARA, or stochastic separability (Karagiannis, 1997a). The latter implies $w_1/w_2 = f_1/f_2 = h_1/h_2$, which is true as long as all inputs have similar risk characteristics.

COROLLARY 1: *Under multiplicative production uncertainty and stochastic complementarity, an output subsidy results in an increase of a risk-increasing input as long as the coefficient of relative risk aversion is less than one.*

PROOF: Given that f(x)=h(x), the bracketed term is reduced to $f_i Eu'(\bullet)\varepsilon + fEu''(\bullet)\pi_i \varepsilon$, which is positive if relative risk aversion coefficient is less than one (Hiebert, 1982). Moreover, cost minimization is ensured under multiplicative production uncertainty (Karagiannis, 1997a) and thus, the fourth term in both (3a) and (3b) is negative.

QED

Relative risk aversion implies that the disutility of proportional risk increases with profit, since an increase in profit increases the dispersion of risk (Hanson and Menezes, 1978). In other words, the condition that relative risk aversion is less than one can be interpreted as implying that the effect of the mean always dominates the effect of risk (variability) (Hiebert, 1982).

COROLLARY 2: *Under additive production uncertainty and stochastic complementarity an output subsidy results in an increase of all inputs.*

PROOF: Given that $\pi_i = 0$, $h_i = 0$ and cost minimization, which is ensured under additive production uncertainty (Karagiannis, 1997a), the right-hand side of (3a) and (3b) collapses to their last term, which is positive if $F_{12} > 0$.

QED

The corresponsibility levy scheme implemented in the CAP is usually coupled with a predetermined maximum quantity guaranteed. Up to the average of last years output, which determines the maximum quantity guaranteed, farmers receive a price including a subsidy, most often a deficiency payment. If output produced exceed the level of maximum quantity guaranteed a levy is charged reducing the amount of deficiency payment. Thus, in terms of our model, a levy is charged if $\varepsilon > 0$; otherwise the corresponsibility levy scheme is inefficient.

The corresponsibility levy scheme operates more or less as a tax that reduces the level of production subsidy. In modeling the corresponsibility levy within the above framework, it is thus clear that it works in an opposite to production subsidy direction. Hence, all algebraic derivations are similar as the above, but comparative static results work in the opposite way.

Policy Implications

In order to derive some results concerning the environmental impact of price support under conditions of production uncertainty, the above mentioned theoretical results should be coupled with empirical evidence about the risk characteristics of fertilizer and pesticide, as well as information about technical interrelationships among inputs (substitutes/complements). It is obvious that these relations may differ across countries, regions and crops due to different weather and soil conditions, water requirements and availability, level of technology etc. Thus, conclusive results cannot be draw by relying solely on the derived comparative static results.

Unfortunately, there are few empirical studies in this field; more importantly, the existing studies reach different results concerning the risk-input relation of pesticides. Leathers and Quiggin (1991) considered pesticide as a risk-reducing input, while Pannell (1991), and Horowitz and Lichtenberg (1994) raised some concerns about its risk nature. Recent empirical evidence by Gotsch and Regev (1996), and Regev, Gotsch and Rieder (1997) show that fungicides behave as a risk-increasing input in Swiss wheat production. On the other hand, Saha, Shumway and Talpaz (1994) found materials (machinery, fertilizer, pesticides and seed) to be risk-reducing while land and structure to be risk-increasing in Kansas wheat production. Nitrogen and phosphorus have been found to be risk-increasing while potassium to be risk-reducing (Love and Buccola, 1991; Regev, Gotsch, and Rieder, 1997).

TABLE 5. 2: Summary of comparative static results

	Risk-reducing Inputs	Risk-increasing Inputs	Conditions
General form of uncertainty	+	?	IPRRA
Multiplicative uncertainty		+	RRA<1
Additive uncertainty		+	-

Based on the above empirical findings and the comparative static results summarized in Table 5.2, it may be argued that increases in price support result in an increase of fertilizer use as long a production uncertainty has a multiplicative or an additive form. Following Love and Buccola (1991) and Regev, Gotsch, and Rieder (1997), this is more likely to occur for nitrogen and phosphorus. On the other hand, if pesticides are considered as a risk-reducing input, an increase in price support causes pesticide use to increase, while the opposite is true for an increase in the level of the corresponsibility levy. Similar results hold for potassium, which in some empirical studies found to be risk-reducing.

For Greek agriculture, in particular, fertilizer and pesticides use per unit of land has steadily increased since the early 1960s, with the most considerable changes taking place after Greece's accession to the EU in 1981. From the early 1960s to the early 1970s, fertilizer use increased in spite of declining fertilizer/ crop output price, while since the early 1980s this trend has continued even though fertilizer/crop output price has been increasing (Lekakis and Kousis, 1994). Pesticide use, on the other hand, has been increasing despite the fact that its relative price has increased. These trends should be explained in the light of demand elasticity estimates for fertilizer and pesticides, but such empirical evidence for Greek agriculture is only available for fertilizer.

A recent study by Mergos and Stoforos (1997), partially supports our theoretical findings, as long as fertilizer is considered to be a risk-increasing input and production uncertainty has either a multiplicative or an additive form. Nevertheless, the price of pesticides has not separately been included in fertilizer demand and hence, their technical interrelationship is unknown. According to their results, the elasticity of fertilizer demand with respect to crop price is, in absolute terms, much greater than the own-price elasticity. This implies that fertilizer use has been affected more by crop output prices than fertilizer price. Thus, for Greek agriculture, higher price support seems to have a greater influence on fertilizer use than does a reduction of its price.

References

Abler, D. G. and J. S. Shortle, 1992. «Environmental and farm commodity policy linkages in the US and the EC,» *European Review of Agricultural Economics*, 19: 197-217.

Beopoulos, N. and D. Skuras, 1997. «Agriculture and the Greek rural environment,» *Sociologia Ruralis*, 37: 255-269.

Brouwer, F. M. and S. van Berkum, 1997. *CAP and Environment in the European Union*, Wangeningen Press, Wangeningen.

deWit, C. T., 1988. «Environmental impact of the CAP,» *European Review of Agricultural Economics*, 15: 283-296.

Gotsch, N. and U. Regev, 1996. «Fungicide use under risk in Swiss wheat production,» *Agricultural Economics*, 14: 1-9.

Hanson, D. L. and C. F. Menezes, 1978. «The effect of capital risk on optimal saving decisions,» *Quarterly Journal of Economics*, 92: 653-670.

Hiebert, L. D. 1982. «Production uncertainty and factor demands for the competitive firm: an extension,» *Southern Economic Journal*, 48: 211-215.

Horowitz, J. K. and E. Lichtenberg, 1994. «Risk-reducing and risk-increasing effects of pesticides,» *Journal of Agricultural Economics*, 45: 82-89.

Hrubovcack, J., M. LeBlank, and J. Miranowski, 1990. «Limitations in evaluating environmental and agricultural policy coordination benefits,» *American Economic Review*, 80: 208-212.

Just, R. E. and J. M. Antle, 1990. «Interactions between agricultural and environmental policies: a conceptual framework,» *American Economic Review*, 80: 197-202.

Just, R. E., and R. D. Pope, 1978. «Stochastic specification of production functions and economic implications,» *Journal of Econometrics*, 7: 67-86.

Karagiannis, G. 1997b. «Interaction between agricultural and resource policy: Note,» *American Journal of Agricultural Economics*, (forthcoming).

Karagiannis, G. 1997a. «Production uncertainty, risk aversion and cost minimization,» *Greek Economic Review*, (forthcoming).

Karagiannis, G. and R. Gray, 1996. «The LeChatelier principle under production uncertainty,» *Scandinavian Journal of Economics*, 98: 453-460.

Leathers, W. D. and J. C. Quiggin, 1991. «Interactions between agricultural and resource policy: the importance of attitudes toward risk,» *American Journal of Agricultural Economics*, 73: 757-764.

Lekakis, J. N. and M. Kousis, 1994. «Agriculture and the environment in Greece, » *Progress in Rural Policy and Planning*, 4: 177-207.

Lewandrowski, J., J. Tobey, and Z. Cook, 1997. «The Interface between agricultural assistance and the environment chemical fertiliser consumption and area expansion,» *Land Economics*, 73: 404-27.

Love, H. A. and S. T. Buccola, 1991. «Joint risk preference-technology estimation with a primal system,» *American Journal of Agricultural Economics*, 73: 765-774.

Magnusson, G. 1969. *Production under Risk: A Theoretical Study*. Uppsala: Almqvist & Wiksells Boktryckeri AB.

Menezes, C. F. and D. L. Hanson, 1970. «On the theory of risk aversion,» *International Economic Review*, 11: 481-487.

Mergos, G. J. and C. E. Stoforos, 1997. «Fertilizer demand in Greece,» *Agricultural Economics*, 16: 227-235.

Murphy, J. A., W. H. Furtan, and A. Schmitz, 1993. «The gains from agricultural research under distorted trade,» *Journal of Public Economics*, 51: 161-172.

Pannell, D. J. 1991. «Pests and pesticides, risk and risk aversion,» *Agricultural Economics*, 5: 361-383.

Pope, R. D. and R. A. Kramer, 1979. «Production uncertainty and factor demands for the competitive firm,» *Southern Economic Journal*, 45: 489-501.

Regev, U., N. Gotsch, and P. Rieder, 1997. «Are fungicides, nitrogen and plant growth regulators risk-reducing? Empirical evidence from Swiss wheat production,» *Journal of Agricultural Economics*, 48: 167-178.

Saha, A., C. R. Shumway, and H. Talpaz, 1994. «Joint estimation of risk preference structure and technology using expo-power utility,» *American Journal of Agricultural Economics*, 76: 173-184.

Zeckhauser, R. and E. Keeler, 1970. «Another type of risk aversion,» *Econometrica*, 38: 661-665.

Chapter 6

Supply Response for Soil Erosive and Chemical Intensive Crops

Joseph N. Lekakis

Introduction

It is widely recognized that the Common Agricultural Policy (CAP) has contributed towards ecological disturbances in Europe. Continued farmer support and high producer prices have led to surplus production and the intensive use of fertilizers and pesticides. This is a major argument advanced by free market optimists for conservation through liberalized agricultural policies (e.g, Anderson and Bluckhurst, 1992). Their contention is that elimination of farmer support will *a priori* be beneficial for the environment. The logic appears simple. A decisive reduction in producer prices could affect returns to agriculture dramatically. Faced with lower marginal revenues, farmers would have to cutback on the use of environmentally damaging inputs. For some liberalizers, additional environmental controls may be necessary to mitigate problems of agricultural pollution (Ford-Runge, 1994).

Liberalizers' views are subject to multiple criticism. First, freer trade does not necessarily imply lower agricultural prices. These are volatile in a free market, considering also the world population growth, newly emerging markets, and a slowdown in productivity growth (Beaumond and Barnett, forthcoming). Thus, removal of farmer support may suppress prices only temporarily. Product differentiation and aggressive marketing might boost production. An improvement of the cotton seed propagation system, for example, could produce Greek varieties of superior fiber quality which can compete successfully in foreign markets, with increased production as a result. Second, environmentalists maintain that the mere reduction of producer prices due to the removal of farmer support, could generate a variety of effects, some of which may be detrimental to the environment (Haigh and Grove-White, 1995).

The 1992 CAP reform, which instituted a lower producer price regime, has been credited by the Commission's «Agenda 2000» for creating some favorable environmental impacts due to an overall reduction in the amounts of agrichemicals used in cereals production during the last decade. However, as recent research indicates, CAP may have been only one of the many factors which have slightly influenced agrichemicals consumption rates in the EU. In addition, while production and environmental impacts are location-specific, lower prices are seen to have complex impacts on the agricultural system and the natural environment (Brouwer and van Berkum, forthcoming). Finally, as the European Commission's Deputy Director General declared in a recent forum in Washington, D.C., there is no evidence that price cuts on their own have led to less intensive production; fertilizer use certainly fell, but this was largely a result of the 'set-aside' program (Roberts, 1997).

With specific reference to Greece, Lekakis (forthcoming) attempted a measurement of agrichemicals productivity in crop production in Greece during the period 1971-1995. Within the limitations of a transcendental logarithmic unit cost function, he found declining productivity for both fertilizers and pesticides. The productivity of fertilizers only slightly reverses its trend after the year 1990, possibly due to the abolition of input subsidies. The productivity of pesticides continues is downward slope throughout the studied period. These findings are by no means applicable to specific crops or regional production processes. In a recent study, Mergos and Stoforos (1997b) conclude that output price reductions for the entire agricultural sector in Greece have been more effective in controlling fertilizer demand than would increasing fertilizer prices due to, say, a fertilizer tax policy. The study does not includes an analysis of pesticide demand.

The main assumption of liberalizers which is considered at fault is the equation of environmental problems solely with the intensive use of agrichemicals, ignoring the implications for wildlife habitat and rural landscapes of sudden cutbacks in farmer support (Potter, forthcoming). On the one hand, abrupt price reductions leading to farm business failures could end in a redistribution of the farms and to dramatic changes in the rural ecology and landscape. On the other hand, given the diversity of farming conditions, the environmental impacts will be neither even nor regular. Responding to price cuts and lower profit margins, farmers may elect to suppress labor-intensive production costs devoted to on-farm conservation. Also, possible are a relocation of production and a loss of high natural value farming systems, especially in the Southern EU countries.

Finally, even within their own hypotheses, liberalizers tend to falsely ignore the recent literature on producer behavior as regards agrichemicals. Lower producer prices imply relatively higher input costs. However, while the elasticity of demand for fertilizers is very small, that for pesticides is close to zero (e.g. Mergos, 1991; Lekakis and Kousis, 1994). For pesticides, especially the intuitive explanation of close to zero elasticities, is that farmers have become dependent on them. As pest resistance to chemicals increases, farmers turn to new more expensive chemicals. This in fact may have been the reason for the continuing decrease of pesticides productivity during the period following the abolition of input subsidies in the EU (Lekakis, forthcoming). Finally, the findings of the Mergos and Stoforos (1997b) study would imply that the impact of supplementary environmental controls, which liberalizers consider necessary in order to mitigate problems of agricultural pollution, would be less decisive if those controls are price-based.

From an agri-environmental policy perspective, it matters to know *a priori* the farmers' supply response to changing prices, especially those who produce soil erosive and chemicals intensive crops. Critics of freer trade have argued that market optimists ignore the possibility of a perverse supply response on the part of some farmers. That is, farmers do not behave as rational producers, and, in an attempt to sustain their gross earnings, they may actually boost production volumes (Potter, forthcoming). At any rate, measuring supply response could give an indication of the impact on the environment anticipated by the liberalizers.

This chapter attempts an empirical estimation of supply response for crops whose production is associated with potential environmental offenses. Table 6.1. shows various crops along with their erosion potential and chemical intensity, listed in the international literature. Of the first five crops included in Table 6.1., Greece produces very small quantities of soybeans, grain sorghum, and rice. Therefore, the analysis will resort to mainly cotton and corn.

Theoretically, since farms may be operating as multi-product enterprises, the analysis of supply response should take into account supply functions of more than one output and input demand functions, which requires estimating simultaneous equation systems (Mergos, 1991). This provides not only more 'accurate' crop own price elasticities, but also cross price elasticities. Mergos' study is the only one known in Greece to have yielded empirical results on this front, using a multi-output profit function approach. However, the analysis is restricted to aggregate crop and livestock supply response. At the empirical level,

Hallam and Zanoli (1993) provide the latest techniques on error correction models regarding supply response. Finally, Mergos and Stoforos (1997a), using the APAS (Agricultural Policy Analysis Simulator) model provide short and long run supply elasticities for several crops including cotton and corn.

Focusing on cotton and corn, experts maintain that they are 'competitive' crops. When the price of cotton goes up relative to the price of corn, farmers plant cotton instead of corn and vice versa. In some areas, both corn and cotton are also 'competitive' to irrigated tobacco and sugar beets, which are also demanding in terms of plant protection (Beopoulos and Skuras, 1997).

The tendency of farmers towards monoculture practices (i.e. their persistence in growing the same crop, or at most one other crop, permanently rather than diversifying to a wider range of crops or applying crop rotation techniques) may provide ground for not using simultaneous equation systems in modeling production. That is, crops may be viewed as industries, wherein the decision making is minimally affected by production decisions made outside the industry. Thus, supply functions based on a variant of the well known Nerlovian approach are still worth estimating, despite the theoretical

TABLE 6. 1: Crops, erosion potential, and chemical intensity

Crops	Relative Erosiveness	Relative Fertilizer Requirements	Relative Pesticide Requirements
Cotton	Most Erosive	High	High
Corn	Moderately erosive	Highest	High
Soybeans	Most erosive	Low	Moderate
Grain Sorghum	Moderately erosive	High	Moderate
Rice	Less erosive	Moderate	High
Wheat & Barley	Less erosive	Moderate	Low

Adapted from: (Reichelderfer, 1990).

and empirical weaknesses associated with them. This approach has been particularly popular among agricultural economists in Greece, and has been widely used in agricultural supply modeling (e.g., Pavlopoulos, 1967; Zanias 1981; Baltas, 1987; Apostolou and Varelas, 1987; Lianos and Rizopoulos 1988).

An Empirical Estimation of Supply Response

Production theory requires that the optimal (equilibrium) supply Y_t^* of a product is a function of the producer price and other exogenous factors such as technological change T, i.e. $Y_t^*=s(p_t,T)$. Assuming a linear stochastic specification of the supply function, equilibrium supply Y^* is written as,

$$Y_t^* = a_0 + a_1 p_t + a_2 T_t + v_t \qquad (1)$$

The Nerlovian approach is based on the proposition that quasi-fixed factors of production and other reasons (e.g. weather) prevent the supply of a crop from reaching its equilibrium level within a time period. Instead, supply adjusts towards the long-run equilibrium supply via a partial adjustment scheme of the type,

$$Y_t - Y_{t-1} = \lambda\left(Y_t^* - Y_{t-1}\right) + u_t \quad 0 \le \lambda \le 1 \qquad (2)$$

where the coefficient λ represents the proportion of the supply's adjustment towards the equilibrium level which occurs within one time period, and u_t denotes a random disturbance term. Combining the partial adjustment process described by (2) and the linear equilibrium supply function (1) finally yields the estimable supply function,[1]

$$Y_t = b_0 + b_1 p_{t-1} + b_2 Y_{t-1} + b_3 T_t + \varepsilon_t \qquad (3)$$

where $b_0 = a_0\lambda, \quad b_1 = a_1\lambda, \quad b_2 = (1-\lambda), \quad b_3 = a_3\lambda$.

Supply equations of the general type described in (3) are estimated below for the four examined crops using annual data for the period 1961-1995. Data on output and producer prices were provided by the Greek Ministry of Agriculture; a general price index of crop production was computed from data published by the National Statistical Service of Greece (NSSG). To specify the empirical form of the supply equations, alternative price series were used including the price p_i actually received by producers of the crop i, the price ratio (p_i/p_j) of the producer price to the price of the most competitive crop p_j and the ratio (p_i/P) of the producer price to the general price index of crop production (Baltas, 1987). Based on preliminary estimation of alternative specifications the following estimable supply equations were chosen:

$$Cotton \quad Y_t = b_C + b_{CC}\left(\frac{p_C}{p_M}\right)_{t-1} + \theta_C Y_{t-1} + v_t \qquad (3.1)$$

$$Tobacco \quad Y_t = b_T + b_{TT}\left(\frac{p_T}{p_M}\right)_{t-1} + \theta_M Y_{t-1} + e_t \qquad (3.2)$$

$$Corn \quad Y_t = b_M + b_{MM}\left(\frac{p_M}{P}\right)_{t-1} + \theta_M Y_{t-1} + \gamma T + \delta D + \omega_t \qquad (3.3)$$

$$Sugarbeets \quad Y_t = b_B + b_{BB}\left(\frac{p_B}{P}\right)_{t-1} + \theta_B Y_{t-1} + \phi T + \varepsilon_t \qquad (3.4)$$

where p_C, p_T, p_M, p_B are the producer prices for cotton, tobacco,[2] corn, and sugar beets; P is the general price index of crop production (1970=100); T is a series taking the values 1, 2,... to account for technical progress; D is a dummy variable taking the value 0 prior to 1981 and 1 afterwards to allow for effects due the entry of Greece into the EU.

Equations (3.1) through (3.4) are not jointly specified and therefore, in principle, they may be estimated separately via OLS. However, it is plausible that random shocks in the supply of one crop may affect the supply of other crops although their production processes are not directly related. If the disturbances of the equations are contemporaneously correlated then more efficient parameter estimates can be obtained by estimating these equations jointly using Zellner's method of seemingly unrelated regressions, SUR (Judge *et al*, 1988).

To allow for the possible existence of contemporaneous correlation among the disturbance terms of the examined supply functions, they were jointly estimated via SUR. An additional complication, however, is that in all four equations the respective lagged dependent variable appears among the regressors. This renders the Durbin-Watson statistic meaningless for testing the existence of serial correlation. To cope with this, the equation set was estimated via SUR assuming first and second order serial correlation in the error terms. Results indicated the existence of second order serial correlation only in the cotton supply equation (first and second order correlation estimates in the rest of the equations were statistically insignificant at any conventional level). Thus the supply equations were estimated as a SUR system with the cotton supply equation corrected for second order autocorrelation; the obtained parameter estimates are shown in Table 6.2.

In addition to the linear specification of the supply equations, a log-linear specification was also considered and estimated as a SUR system. The main difference between a linear and log linear specification is that the former allows elasticity point estimates to vary across the sample, while the latter implicitly imposes a constant elasticity structure on the supply functions. In the case of logarithmic specification, no serial correlation was detected in any of the supply equations; the obtained estimates are also shown in Table 6.2.

TABLE 6. 2: Estimation results

	Linear specification		Log - linear specification	
Coefficient	Estimate	Std - error	Estimate	Std - error
b_C	-191.53	48.70	0.808	0.369
b_{CC}	50.46	13.29	0.565	0.138
θ_C	0.889	0.07	0.742	0.083
b_T	23.41	15.60	1.432	0.495
b_{TT}	2.01	0.819	0.310	0.113
θ_T	0.52	0.120	0.517	0.113
b_M	-1282.3	623.2	0.206	0.874
b_{MM}	483.83	211.03	0.678	0.280
θ_M	0.374	0.173	0.844	0.163
γ	41.54	12.59	0.021	0.011
δ	481.12	166.56	0.034	0.133
b_B	-1777.54	978.71	3.942	0.550
b_{BB}	4343.35	1791.66	1.106	0.420
θ_B	0.39	0.14	0.526	0.704
ϕ	62.97	16.18	0.031	0.008
Cotton:	$R^2 = 0.966$	DW = 2.098	$R^2 = 0.92$	DW = 2.14
Tobacco:	$R^2 = 0.540$	DW = 2.076	$R^2 = 0.57$	DW = 1.99
Corn:	$R^2 = 0.954$	DW = 2.174	$R^2 = 0.97$	DW = 1.67
Sugarbeets:	$R^2 = 0.709$	DW = 2.079	$R^2 = 0.89$	DW = 2.18

Inspection of the estimation results reveals that with the exception of tobacco, the explanatory power of both the linear and log-linear specifications (as expressed by the respective adjusted R^2) is fairly high. In the linear specification, all estimated coefficients are significant, at least at the 5 percent level, except for the intercept terms a_T, a_B. Technical progress is a significant determinant of corn and sugar beet supply, while the entrance of Greece into the EU implies a significant impact on the supply of corn. The Nerlovian adjustment coefficient is highest in the case of cotton, implying almost complete adjustment within one period; the same coefficient is fairly low in the cases of corn and sugar beets indicating a much slower adjustment of supply towards equilibrium levels.

Turning to the log linear specification, it is observed that here too all coefficients are statistically different from zero, at least at the 5 percent level, with the exception of the intercept a_C and the coefficients of technical progress and 'EU entrance' in the case of corn. Regarding the estimated adjusted coefficients, they are fairly close to those of the linear specification, except for the adjustment coefficient of corn, which here is more than twice the respective estimate in the linear specification.

The degree of response of the examined supply functions to price changes may be empirically investigated by inspecting the elasticity estimates presented in Table 4.3. The left side of the table shows short run own price elasticities over time based on the linear specification of the supply equations. The respective long run elasticities appear on the right side. Finally, same elasticity estimates based on the log linear specification appear in the lower part of the table.

The obtained elasticity estimates are fairly comparable to those found by other researchers. Lianos and Rizopoulos (1988) and Zanias (1981) found the short run elasticity of supply of cotton to be 0.703 and 0.65 for the linear specification, respectively. Our long run elasticity of supply of cotton is closer to those found by Zanias (3.1 to 3.53). Moreover, Baltas (1987) had estimated the short and long run supply elasticity for corn as 0.5 and 2.8 for a log linear specification, respectively. Finally, the short run elasticity estimates are also comparable to those of the APAS model by Mergos and Stoforos (1997a). The long run estimates of the latter study are much lower than those reported by this chapter. Examination of the reported short run elasticities based on a linear specification of the supply equations indicates the following:

Short run elasticities for cotton, corn, and sugar beets show a decreasing trend over time, while the short run elasticity of tobacco is fairly stable over the examined period. In other words, cotton, corn,

TABLE 6. 3 : Average own price elasticities of supply

Linear specification

	Short-run own price elasticities				Long-run own price elasticities			
Period	Cotton	Tobacco	Corn	S-beets	Cotton	Tobacco	Corn	S-beets
1965-71	0.57	0.26	3.31	2.42	5.13	0.53	5.29	3.97
1972-81	0.50	0.29	1.67	0.97	4.50	0.61	2.66	1.58
1982-91	0.45	0.30	0.48	0.73	4.05	0.63	0.77	1.19
1992-95	0.31	0.33	0.36	0.58	2.79	0.68	0.57	0.95
1965-95	0.48	0.29	1.49	1.17	4.32	0.61	2.38	1.92

Log - linear specification

	Short-run own price elasticities				Long-run own price elasticities			
Period	Cotton	Tobacco	Corn	S-beets	Cotton	Tobacco	Corn	S-beets
1962-95	0.56	0.31	0.68	1.10	2.19	0.64	4.35	2.32

and sugar beets producers appear less responsive to price changes of their produce over time.

On average, the short-run responses of cotton and tobacco supply are rather low; a 1 percent decrease in the relative price of cotton (or tobacco) to corn reduces the respective supply by only 0.48 or 0.29 percent, respectively. By contrast, corn, and sugar beet supply are fairly responsive to price changes.

Comparing short run elasticities of supply in various time periods, it is noticed that they are lower in the post MacSharry reform period of 1992-95 than in 1965-1971. A decrease in the relative prices of all four crops by one would reduce their respective supply by a factor between 0.31 and 0.58. The low supply elasticities might be attributed, among others, to the guaranteed prices under the CAP.

Adjustment towards equilibrium supply levels (as inferred from the computed long run elasticities) is fairly rapid for cotton followed by corn and sugar beets; it is however rather slow for tobacco. Almost analogous conclusions may be drawn from the study of the elasticity estimates based on the log linear specification.

Conclusions

The empirical analysis of supply response for the examined crops suggests that considerable reductions in their supply levels would require deep cuts in the prices received by farmers. At least in the short run, reductions in the current price support schemes do not guarantee appreciable improvements in environmental conditions.

If cotton and corn are truly 'competitive' in nature, a change in the producer price of one relative to the price of the other, would simply shift agricultural production from a relatively more erosive crop to a relatively agrichemicals intensive crop and vice versa. The higher short run elasticity of corn supply, compared to cotton, does not necessarily imply that a drop in the relative price of the former, leading to a possible reduction in the amounts of fertilizers used, will enhance environmental conditions. The switch to cotton production would still require high fertilizer application rates, and the environmental effects of agrichemical pollution are known to be cumulative and long-lived. Price reductions in industrial crops such as corn and sugar beets would lead to a meager switch of production to non-industrial, less environmentally demanding crops. In general, it appears that crop production is locked in monocultures and minor changes across crops would not result in decisive environmental benefits.

However, the environmental implications stemming from the empirical results of supply response should not be exaggerated. It should be borne in mind that the preceding empirical framework is based on the well-known pressumptions of the standard economic theory concerning the uniqueness of the underlying mode of production and the farmer's decision criteria. This definitely leaves out of the analysis worthy alternatives. For instance, utilizing alternative modes of production (e.g. organic farming) could induce farmers to simultaneously reduce surplus produce, lower costs, and enhance environmental conditions.

Acknowledgement: Many thanks to Dr. Christos Pantzios for the 'number crunching' and to Professor George Mergos for his comments on an earlier draft. Naturally, any remaining errors are solely mine. Special thanks to Mr. Evangelos Kazazis, currently with the Greek Ministry of Agriculture, Agricultural Statistics Division, in Athens, for the data used in the calculations, and for our long discussions over the phone.

Notes

[1] The output partial adjustment process given in (2) may also be combined with price uncertainty by introducing a process via which the producers form their expectations about prices (an adaptive expectation scheme). However, in regimes such as the one considered here, prices are preset and announced by the government. Thus, price uncertainty may be left out of the analysis.

[2] The producer price of tobacco was computed as a production weighted average of producer prices for Barley, Virginia, and oriental type tobacco.

References

Anderson, K. and R. Blackhurst (eds), 1992. *The Greening of World Trade Issues*, London: Harvester Wheatsheaf.

Apostolou, N. and E. Varelas, 1987. *Supply Response Functions for Selected Agricultural Commodities*, Studies for the Agricultural Economy No 25, Athens: Agricultural Bank of Greece (in Greek).

Baltas, N. C. 1987. «Supply Response for Greek Cereals,» *European Review of Agricultural Economics*, 14(2): 195-220.

Beaumond, H-C. and A. Barnett, (forthcoming). «The Opportunities and Risks for the European Rural Environment Associated with the Liberalization of Agricultural Trade,» in M. Redclift, J. N. Lekakis, and G. P. Zanias (eds), *Agriculture and World Trade Liberalization: Socio-environmental Perspectives on the Common Agricultural Policy*, Wallinford, UK: CAB International.

Beopoulos N. and D. Skuras, 1997, «Agriculture and the Greek Rural Environment,» *Sociologia Ruralis*, 37(2): 255-269.

Brouwer, F. and S. van Berkum, forthcoming. «Effects of CAP Reform on the Environment in the EU,» in Antle, J. M, J. N. Lekakis, and G. P. Zanias (eds), *Agriculture, Trade, and the Environment: The Impact of Liberalization on Sustainable Development*, Cheltenham, UK: Edward Elgar.

Ford Runge, C. 1994. *Freer Trade, Protected Environment: Balancing Trade Liberalization and Environmental Interests*, New York: Council of Foreign Relations Press.

Haigh, N. and R. Grove-White, 1995. «Introduction and Observations,» in D. Baldock and D, Conder (eds), Can the CAP fit the Environment? Proceedings of a seminar, London: CPRE and Institute for European Environmental Policy.

Hallam, D. and R. Zanoli, 1993. «Error Correction Models and Agricultural Supply Response, *European Review of Agricultural Economics*, 20: 151-166.

Judge, G. G, R. Carter Hill, W. E. Griffiths, H. Lutkepohl, and T. C. Lee, (1988). *Introduction to the Theory and Practice of Econometrics*, 2nd Edition, Johh Wiley & Sons Inc.

Lekakis, J. N. forthcoming. «The Productivity of Agrochemicals in Greece,» in Antle, J. M, J. N. Lekakis, and G. P. Zanias (eds), *Agriculture, Trade, and the Environment: The Impact of Liberalization on Sustainable Development*, Cheltenham, UK: Edward Elgar.

Lekakis, J. N. and M. Kousis, 1994 «Agriculture and the Environment in Greece,» *Progress in Rural Policy and Planning*, 4: 177-207.

Lianos Th. P. and G. Rizopoulos, (1988). «Estimation of Social Welfare Weights in agricultural Policy: The Case of Greek Cotton,» *Journal of Agricultural Economics*, 38: 303-313.

Mergos, G. J. 1991. *Output Supply and Input Demand in Greek Agriculture: A Multi-Output Profit Function Approach*, Athens: Center of Planning & Economic Research.

Mergos G. J. and C. E. Stoforos, 1997a. «Supply Response in Greek Agriculture: A Cointegration Analysis,» Department of Economics, University of Athens.

Mergos G. J. and C. E. Stoforos, 1997b. «Fertilizer Demand in Greece,» *Agricultural Economics*, 16: 227-235.

Pavlopoulos, P. 1967. *Supply Functions of Agricultural Products: A Quantitative Investigation*, Athens: Center of Planning & Economic Research.

Roberts, D. F. 1997. «The Choices Facing the European Community in the Light of the Progress Towards More Open Markets,» in USDA, *Agricultural Outlook Forum*, 1997 Proceedings, Washington D.C., April, 24-25 (NTIS, WAOB-YCON-97).

Potter, C. (forthcoming). «Agricultural Liberalization and the 'Double Dividend',» in M. Redclift, J. N. Lekakis, and G. P. Zanias (eds), *Agriculture and World Trade Liberalization: Socio-environmental Perspectives on the Common Agricultural Policy*, Wallinford, UK: CAB International.

Reichelderfer, K. (1990). «Environmental Protection and Agricultural Support: Are Trade-Offs Necessary?» in K. Allen (ed) *Agricultural Policies in a New Decade*, Washington, DC: Resources for the Future.

Zanias, G. P. 1981. «An Estimation of the Supply Response of Cotton in Greece from 1950-1979,» *Oxford Agrarian Studies*, 10: 196-211.

Chapter 7

Economic Implications of Fertilizer Limitation Policies

Evangelos Papanagiotou

Introduction

In recent years, Western European Agriculture has been focusing on two main problems, namely, food surpluses and agricultural pollution. Fertilizers play a very important role in both of the aforementioned problems by increasing crop yields and contributing to the pollution of surface and ground water supplies, to air pollution, and to land degradation (De Haen, 1982; Simonis and Setatou, 1991; Simonis, 1992a).

Chemical fertilizers account for only about 6% (1-12%) of the cost of crop production in Greece (Kitsopanidis *et al,* 1983; Velemis, 1991), while they contribute to the crop yields by about 30-40% (Simonis 1992b). In general, fertilizers are only responsible for 3.5% of the nitrate pollution by (Simonis, 1992b). The problem of nitrate pollution is more evident at the local level. For example, in Germany and in Denmark, 6% and 2% respectively of drinking water contains nitrates over the limits, due to farm activities (Conway, 1990). This happens because 10-50% of the applied fertilizer is lost into the environment. According to De Haen (1982), Colenbrander estimates this loss to be 43% for arable crops and 34% for permanent grass. It appears, therefore, that there is a close relation between fertilizer use and nitrates in the surface and ground water.

With reference to Greece, scientists warn of the danger from the excessive amounts of nitrogen fertilizer applied to crops (Almaliotis and Paloukis, 1995). Current research on nitrates in underground water (Kritikos and Apostolidis, 1997) finds nitrates over the limits in a number of regions and suggests that the main cause of it is the agricultural activity. Also, research at the local level suggests that the excessive use of nitrogen fertilizer leads to land degradation (Papanagiotou, 1995).

In addition, the work of Mourkidis (1986) indicates that agricultural activity is responsible for about 45% of the nitrate eutrophism.

Therefore, either because of the need to limit surpluses and reduce the cost of the agricultural policy (Marsh, 1987; Weinschenck, 1987), or because of environmental reasons (Papanagiotou and Melfou, 1992; Papanagiotou, 1995), policies aimed at reducing the consumption of certain agricultural inputs -including fertilizers- are currently under consideration in Greece.

The OECD, the EU, and other international organizations have suggested a number of measures to address the problems of production surpluses and agricultural pollution. The measures may be grouped into the following categories: 1) General measures. These include direct measures such as the encouragement of ecological investments, and indirect measures such as the encouragement of extensive farming methods. 2) Economic measures such as those affecting crop prices and agrichemical prices. 3) Legislative measures such as quotas and other restrictions. 4) Voluntary measures to increase chemical efficiency. 5) Education and research. The aforementioned measures may have some negative effects on farmers, manufacturers, and on society as a whole since they may lead to a reduction in crop yields and farm incomes, as well as lower efficiency and competitiveness. The economic effects of these measures are generally difficult to evaluate. The next section provides a brief review of the alternative approaches employed by researchers to quantify the economic consequences of different policy measures.

A Review of Alternative Approaches

An extensive review of past works which have attempted to quantify the economic effects resulting from lower fertilizer consumption regimes is provided by Papanagiotou (1996). A discussion on the methodology and the results of the most representative of those works follows.

McCorriston and Sheldon (1989), employing a theoretical model of the UK fertilizer industry along with simulation techniques calculate the multipliers for a 10% increase of fertilizer price and a 10% decrease of fertilizer quantity, induced by quotas. Alexandratos, Bruinsma, and Stamoulis (1990) use a simple one output-two input model of the agricultural sector (one of the inputs being fertilizer) to calculate the effects of the imposition of a fertilizer tax under alternative policy regimes. De Haen (1982), using experimental results on the ratio between nitrates in the groundwater and nitrogen fertilizer use, and the relation between crop yield and nitrogen applications, estimates the income effect of a decrease in fertilizer use.

England (1986), using a profit maximization condition along with a nitrogen response function, estimates the optimum nitrogen intake for existing prices of a number of grain products. He then changed the fertilizer and product prices and estimated the effects on prices. Possible disadvantages of this work are the use of only one factor to explain yields and the fact that crops are not considered in the context of the total farm system.

A considerable amount of research has been directed toward the estimation of own and cross price elasticities for agrichemicals. Examples can be found in the works of Burrell (1989), and Mergos (1991). Finally, a number of studies, at the farm or regional level, rely on programming techniques such as Partial Budgeting and Linear Programming (Papanagiotou, 1996; Zekri and Herruzo, 1993) to identify the effects of alternative policies on crop mix, farm incomes, and environmental indices.

The aforementioned works are limited by the lack of an underlying aggregate model of the agricultural sector estimated from empirical data. An attempt in this direction has been made in an earlier paper by Papanagiotou and Melfou (1993). The objective of this chapter is to quantify the economic effects resulting from a reduction in fertilizer consumption by completing and extending the work of Papanagiotou and Melfou (1993).

The Theoretical Model and the Data

The theoretical model employed in this work begins with the profit equation,

$$Pr = \Sigma(p_j f(x_i)) - \Sigma(w_i X(w_i))$$

where, Pr is profit, p_j is the price of product j, x_i is the quantity of factor i, w_i is the price of factor i, f is the production function, and X is an input demand function

The output of the agricultural sector is assumed to involve both crop and livestock production. The production inputs can be classified into variable and fixed. The variable inputs include fertilizers, feed stuffs, variable capital for crops and variable capital for livestock. The fixed inputs include land and livestock capital. Land has been converted into a homogeneous input (irrigated acreage), while livestock capital is represented by the number of animal units.

According to the National Statistical Service of Greece, during the last 30 years, the annual fertilizer consumption has increased by 5.4%, while the deflated fertilizer price has decreased by 4.4%. Farm output has increased by 2.5% (2.3% for crops and 3% for livestock), while the deflated output prices have increased by only 0.2% (0.3% for crops and 0% for livestock). It appears, therefore, that: a) there is a negative relationship between fertilizer consumption and its price and a positive relationship between fertilizer consumption and both yield and crop prices; b) there is a positive relationship between agricultural production and fertilizer consumption. The behavior of other intermediate inputs is similar. Consequently, the data are suitable to test production functions and input demand functions for the agricultural sector with emphasis on fertilizer input.

The above observations lead to the construction of the following structural model of the agricultural sector in Greece.

$$Y_1L = F(X_1L, X_3L, T) \quad (1)$$
$$Y_2Z = F(X_2Z, X_4Z, T) \quad (2)$$
$$X_1L = X(P_1S, W_1S, Y_1L, T) \quad (3)$$
$$X_2Z = X(P_2S, W_2S, T) \quad (4)$$
$$X_3L = X(P_1S, W_3S, T) \quad (5)$$
$$X_4Z = X(P_2S, W_4S, T) \quad (6)$$
$$PrS = Y_1{*}P_1S+Y_2{*}P_2S-X_1{*}W_1S-X_2{*}W_2S-X_3{*}W_3S-X_4{*}W_4S \quad (7)$$

where:

PrS is the gross margin deflated by consumer price index

Y_1 is crop production, Y_1L is crop yield per unit of land ;

Y_2 is livestock production, Y_2Z is livestock yield per animal unit;

X_1 is fertilizer used, X_1L is fertilizer used per unit of land;

X_2 is feed stuffs consumed, X_2Z is feed stuffs consumed per unit of animal;

X_3 is variable capital used for crop production, X_3L is variable capital per unit of land;

X_4 is variable capital used for livestock production, X_4Z is variable capital per unit of animal;

P_1S is crop price deflated by the consumer price index, P_2S is livestock production price deflated by the consumer price index;

W_1S is fertilizer price deflated by the consumer price index, W_2S is feed stuffs price deflated by the consumer price index;

W_3S is crop capital price deflated by the consumer price index; and,

W_4S is livestock capital price deflated by the consumer price index, and T a time trend which captures technological changes.

The structural model has been estimated using two different data sets. The first data set comes from the National Accounts of Greece (Ministry of Finance) and covers the period 1961-1990. The second data set comes from the *revised* National Accounts of Greece and covers the period 1973-96. Crop production, livestock production and factor quantities have been used with their respective values at constant prices. Price indices for outputs and inputs have been calculated by dividing their values at current prices by the respective values at constant prices. The number of animal units and the level of cultivated land have been obtained from the National Statistical Service (NSSG 1961-1996).

Empirical Results

The model described by equations (1) through (7) is a simultaneous system. The 2SLS (Two Stages Least Squares) method has been used for estimation in order to account for simultaneity bias, using the TSP (Time Series Program) program (Hall and Hall, 1981; Hall and Schnake, 1987). Both linear and double log models have been estimated. The double log models gave the best results.

The asymptotically efficient parameter estimates along with their corresponding standard errors for the period 1961-90 follow[(1)]

Crop production function

$$(1)\ LY_1L = 5.3669 + \underset{(0.0804)}{0.1619}\ LX_1L + \underset{(0.1577)}{0.2981}\ LX_3L + \underset{(0.0329)}{0.1687}\ D$$

$$R^2 = 0.9116 \qquad DW = 1.9509$$

Livestock production function

$$(2)\ LY_2Z = 1.3553 + \underset{(0.0812)}{0.4492}\ LX_2Z + \underset{(0.0805)}{0.3509}\ LX_4Z$$

$$R^2 = 0.6546 \qquad DW = 1.8379$$

Fertilizer demand function

$$(3)\ LX_1L = -0.8133 + \underset{(0.1825)}{0.4121}\ LP_1S(-1) - \underset{(0.0886)}{0.6504}\ LW_1S + \underset{(0.1925)}{0.6872}\ LY_1L(-1) + \underset{(0.0449)}{0.1071}\ D$$

$$R^2 = 0.9318 \qquad DW = 1.9300$$

Feed stuffs demand function

$$(4)\ LX_2Z = 0.9450 + \underset{(0.5113)}{2.4234}\ LP_2S(-1) - \underset{(0.2741)}{1.6772}\ LW_2S + \underset{(0.1243)}{0.2881}\ D$$

$$R^2 = 0.7053 \qquad DW = 1.5871$$

Crop variable capital demand function

$$(5)\ LX_3L = 4.9125 - \underset{(0.0703)}{0.4575}\ L(W_3S/P_1S) + \underset{(0.0025)}{0.0232}\ T$$

$$R^2 = 0.9884 \qquad DW = 1.6761$$

Livestock variable capital demand function

$$(6)\ LX_4L = -0.2816 - \underset{(0.3155)}{0.6543}\ L(W_4S/P_2S) + \underset{(0.0024)}{0.0413}\ T$$

$$R^2 = 0.9110 \qquad DW = 1.9849$$

---

(1) Notes
* standard-errors in parentheses.
**T in equations (1), (2), (3), (4) is found to be statistically insignificant_and therefore it has been omitted.
*** D represents a dummy variable for the drought years.
**** (-) represents a year lag.
***** Equations (2), (3), (5) and (6) have been corrected for first order serial correlation.

The estimated model fits the data reasonably well. In particular, the included independent variables explain a large part of the variability in the relevant dependent variables (R^2 ranges from 0.65 to 0.99), and the coefficients of the independent variables have the theoretically expected signs. The omission of variables associated with statistically insignificant coefficients has not changed the signs of the remaining coefficients and it has only a marginal impact on R^2s.

Equations (1) and (3) have the demand for fertilizers as an explanatory and a dependent variable, respectively. Both perform well. According to equation (1), the production elasticity of fertilizer is 0.16, which means that an increase in fertilizer consumption by 10% will lead to an increase of yields by 1.6%. The consequences of this admittedly low response of yield to fertilizer could be better explained if one takes into account the contribution of fertilizer in crop production cost (which is estimated at the average level of 6%). When the two estimates are considered together, it appears that a 10% increase of fertilizer consumption will raise total costs by only 0.6% and gross output by 1.6%.

This makes clear why the use of fertilizer is profitable.

Equation (3) gives the fertilizer elasticities with respect to own-price (-0.6505), to crop price (0.4121), and to crop yield (0.6872). One, therefore, may conclude that fertilizer consumption depends firstly on the expected yield, then on the current fertilizer price and lastly on the crop price. These results are very similar to those reported by Mergos (1991).

Asymptotically efficient parameter estimates along with their corresponding standard errors for the Period 1973-96 (Revised National Accounts) follow. Again, the double log forms gave the best results.

Crop production function

$$(1)\ LY_1L = 2.2266 + \underset{(0.0563)}{0.1322}\ LX_1L + \underset{(0.0581)}{0.4149}\ LX_3L + \underset{(0.0459)}{0.1340}\ D_1$$

$$R^2 = 0.7424 \qquad DW = 1.6030$$

Livestock production function

$$(2)\ LY_2Z = 1.9932 + \underset{(0.0812)}{0.2273}\ LX_2Z + \underset{(0.0805)}{0.7578}\ LX_4Z$$

$$R^2 = 0.9948 \qquad DW = 1.8062$$

Fertilizer demand function

$$(3)\ LX_1L = -2.0155 + \underset{(0.1997)}{0.5072}\ LP_1S_{(-1)} - \underset{(0.1725)}{0.4966}\ LW_1S +$$

$$\underset{(0.3533)}{0.53111}\ LY_1L_{(-1)} + \underset{(0.0654)}{0.18092}\ D1$$

$$R^2 = 0.7777 \qquad DW = 1.4182$$

Feed stuffs demand function

$$(4)\ LX_2Z = 3.7484 - \underset{(0.8389)}{1.3061}\ L(W_2S/P_2S_{(-1)}) - \underset{(0.2253)}{0.320}\ D_2$$

$$R^2 = 0.5867 \qquad DW = 1.8728$$

Crop variable capital demand function

$$(5)\ LX_3L = 0.3483 - \underset{(0.0179)}{0.1466}\ L(W_3S/P_1S_{(-1)}) + \underset{(0.0236)}{0.0515}\ D_3$$

$$R^2 = 0.3827 \qquad DW = 1.8167$$

Livestock variable capital demand function

$$(6)\ LX_4L = 1.4449 - 0.1136\ L(W_4S/P_2S_{(-1)}) + 0.0281\ D_3$$
$$(0.0483) \qquad\qquad (0.0329)$$
$$R^2 = 0.4891 \qquad DW = 1.9651^{(2)}$$

(2) Notes:

* standard-errors in parentheses.

**T in equations found to be statistically insignificant, and therefore it has been omitted.

*** D_1, D_2, D_3 represent dummy variables for the drought years, years with radiation problems (1986-1987), and a change in CAP (year 1992) respectively.

**** $_{(-)}$ represents a year lag.

***** All equations but (3) have been corrected for first order serial correlation.

The coefficients of the independent variables have the theoretically expected signs. The coefficients of determination , however, are somehow lower that those of the previous model. This is especially true for equations (4), (5), and (6).

According to this model, the elasticity of production with respect to fertilizer is 0.1322. The own-price elasticity of fertilizer is -0.4966. The elasticity of fertilizer demand with respect to crop price is 0.5072 while its elasticity with respect to yield is 0.5311.

Looking at the results from the two estimated models in a comparative way, we may conclude that:

1) The production elasticity of fertilizers ranges from 0.13 to 0.16, showing a slight tendency to decrease over time, which does not necessarily violate the original assumption of constant elasticity. This verifies results from earlier studies on the topic (Lekakis, 1998).

2) Fertilizer demand elasticities with respect to its own price and with respect to yield show a slight tendency to decrease over time, which does not necessarily violates the assumption of constant elasticity. The fertilizer demand, however, with respect to crop price tends to become more elastic over time. It appears, therefore, that environmental policy measures involving changes in the price of fertilizers will become less effective in the future. In contrast, measures involving changes in crop prices will become more effective.

Simulation and Multipliers Analysis

Simulation has been used to test the dynamic behavior and the forecasting power of the model and to estimate the dynamic multipliers of each endogenous variable with regard to the exogenous variables.

The evaluation of the forecasting ability of the model has been carried out using the criteria presented in Table 7.1. In this table, *r* and *RC* are the correlation and regression coefficients between the actual and the simulated variables respectively. *U* is Theil's inequality coefficient, and *UM, US*, and *UC* are the decomposition of U into its inequality proportions, namely, bias (*UM*), variation (*US*), and co-variation (*UC*). The magnitude of these coefficients suggests that the forecasting performance of the model is highly satisfactory (Theil, 1966).

Table 7. 1: Simulation performance coefficients

Variables	*r*	*RC*	*U*	*UM*	*US*	*UC*
Y_1L	0.94086	1.03918	0.00004	0.00041	0.07710	0.92248
Y_2Z	0.93621	1.05692	0.00203	0.01291	0.10215	0.88495
X_1L	0.97464	1.03173	0.00030	0.00918	0.05952	0.93130
X_2Z	0.83986	1.00000	0.02332	0.00000	0.08704	0.91296
X_3L	0.95339	0.99516	0.00013	0.00299	0.01929	0.97772
X_4Z	0.97738	1.01034	0.02347	0.00472	0.02363	0.97165
Pr	0.99955	1.00351	0.00005	0.03471	0.01637	0.94891

The estimated dynamic multipliers show the effect of an initial unit change in the level of a predetermined exogenous variable (e.g. 10%) upon the endogenous variable and are presented in Table 7.2.

Table 7. 2: Dynamic multipliers

Endogenous	Exogenous variables		
Variables	P_1	X_1	W_1
	-10.00	-10.00	+10.00
Y_1L	-0.88	-	-1.04
X_1L	-5.14	-	-7.27
PrS	-10.17	0.39	-0.39

According to the results of the multiplier analysis, a 10% decrease in crop production prices will lead to a very small change in crop yields (-88%), but the decrease will lead to a considerable reduction in fertilizer consumption (-5.14%) and a substantial change in farm income (-10.17%). An increase in fertilizer price by 10%, however, will have a strong effect on fertilizer consumption (-7.27%), a moderate effect on yields (-1.04%), and a negligible effect on farm income (-0.39%). If these changes, in crop and fertilizer prices, take the form of a refundable tax, 90% and 68% of farm income losses respectively can be refunded.

Conclusions

The aim of this chapter, which extends prior work by Papanagiotou and Melfou (1993), has been to investigate the economic implications of a lower fertilizer consumption regime in Greece. In contrast with earlier works on this topic, we estimate here a model of the farm sector and use the estimates for multiplier analysis. The main conclusions from the empirical analysis are: (1) Policy measures involving changes in the price of fertilizers are likely to become less effective in the future. In contrast, policy measures involving changes in crop prices are likely to become more effective. (2) Taxes on fertilizers will have a rather small effect on farm incomes since an increase in the price of fertilizers by 10% will lead to a loss in farm income of only 0.12%. They will, however, have a considerable effect on fertilizer demand since a 10% increase in the price of fertilizers will lead to a reduction in fertilizer demand by 7.27%. (3) Reductions in crop prices will have a substantial impact on farm income since the elasticity of farm income with respect to crop price is close to one. Crop prices reduction will lead, however, to moderate decrease in fertilizer consumption since the relevant price elasticity of demand is close to 0.5.

References

Alexandratos, N., J. Bruinsma and K. Stamoulis, 1990. «Environmental Restrictions in Agriculture: Impact on Other Policy Objectives,» VIth Congress of European Association of Agricultural Economists, The Hague, The Netherlands.

Allison, J. R. 1990. «Internalizing and Controlling the External Costs of Agricultural Chemical Use,» VIth Congress of European Association of Agricultural Economists, The Hague, The Netherlands.

Almaliotis, D. and H. Paloukis, 1995. «Losses in Nitrogen During the Application of Nitrogen Fertilizer,» *Geotechnical Scientific Issues*, 6(4): 18-26.

Burell, A. 1989. «The Demand for Fertilizer in the United Kingdom,» *Journal of Agricultural Economics*, 40: 1-20.

Conway, A. G. 1990. «A Role for Economic Instruments in Reconciling Agricultural and Environmental in Accordance with the Polluter Pays Principle,» VIth Congress of European Association of Agricultural Economists, The Hague, The Netherlands.

De Haen, H. 1982. «Economic Aspects of Policies to Control Nitrate Contamination Resulting from Agricultural Production,» *European Review of Agricultural Economics*, 9: 443-465.

England, R. A. 1986. «Reducing the Nitrogen Input on Arable Farms,» *Journal of Agricultural Economics*, 37: 13-24.

Hall, B. H. and R. E. Hall, 1981. «Time Series Processor (TSP),» Stanford, USA.

Hall, B. H. and R. Schnake, 1987. «Time Series Processor. TSP International,» California.

Kitsopanidis, G., E. Papanagiotou, M. Martica, A. Psychadakis, and B. Manos, 1983. «The Cost of Crop and Livestock Products in Greece,» Department of Agricultural Economics Research, University of Thessaloniki.

Kritikos, G. and G. Apostolidis, 1997. «Nitrogen - the Big Danger for the Water,» *Ethnos* (daily Athenian Press), 28 September 1997, pp 30-31.

Lekakis, J. N. 1998. «The Productivity of Agrochemicals in Greece», in Antle J. M., J. N. Lekakis, and G. P. Zanias (eds), *Agriculture, Trade, and Environment: The Impact of Liberalization on Sustainable Development*, Cheltenham, UK: Edward Elgar (forthcoming).

Marsh, J. 1987. «Alternative Policies for Agriculture,» *European Review of Agricultural Economics*, 10: 11-21.

McCorriston, S. and I. Sheldon, 1989. «The Welfare Implications of the Nitrogen Limitation Policies,» *Journal of Agricultural Economics*, 42: 143-151.

Mergos, G. J. 1991. *Output Supply and Input Demand in Greek Agriculture: A Multi-Output Profit Function Approach*, Athens: Center of Planning and Economic Research, Studies No. 6.

Ministry of Finance, 1961-90. *National Agricultural Accounts*, Athens.

Ministry of Finance, 1973-96. *Revised National Agricultural Accounts*, Athens.

Mourkidis, G. 1986. «The Lakes in Northern Greece,» *Scientific Annals*, School of Agriculture, University of Thessaloniki, Vol. 25.

NSSG (National Statistical Service of Greece), 1961-96. *Statistical Yearbooks of Greece*, Athens.

NSSG, 1961-96. *Agricultural Statistics*, Athens.

Papanagiotou, E. 1995. «Environment and Agriculture - New Directions and Research,» paper presented at a Conference of the Geotechnical Chamber of Greece, Athens 10-12 December.

Papanagiotou, E. 1996. «Negative Effects from the Adoption of Environmental Measures -Problems in Estimation,» 3rd Congress of the Hellenic Society of Agricultural Economists, Athens, Proceedings pp 225-232.

Papanagiotou, E. and K. Melfou, 1992. «An Economic Evaluation of the Environmental Consequences of Agriculture: Contemporary Developments on Research Methodology,» 2nd Congress of the Hellenic Society of Agricultural Economists, Thessaloniki, 11-12 December, p. 21.

Papanagiotou, E. and K. Melfou, 1993. «Some Implications of Fertilizer limitation Policies,» VIIth Congress of European Association of Agricultural Economists, Stressa Italy.

Simonis, A. D. and H. B. Setatou, 1991. «Nitrogen Use Efficiency - Factors Affecting it and Means of its Improvement,» *Geoponica*, 335: 136-146.

Simonis, A. D. 1992a. «Fertilizer and Environment,» *Geoponica*, 339: 74-89.

Simonis, A. D. 1992b. «The Importance of Fertilizers in Agriculture,» *Geoponica*, 341: 135-149

Theil, H. 1966. *Applied Economic Forecasting*, Amsterdam: North-Holland.

Velemis, D. 1991. «Fertilizers - Present Situation and Prospects,» paper presented in a two day conference of the Hellenic Soil Science Society, Thessaloniki 1989.

Weinschenck, G. 1987. «The Economic or the Ecological Way? Basic Alternatives for the EC's Agricultural Policy,» *European Review of Agricultural Economics*, 10: 49-60.

Zekri, S. and A. C. Herruzo, 1993. «Complementary Instruments to EEC Nitrogen Policy on Non-Sensitive Areas: A Case Study in Southern Spain,» VIIth Congress of the European Association of Agricultural Economists, Stressa Italy.

Chapter 8

Towards a More Competitive and Sustainable Cotton Production

Stella Galanopoulou-Sendouca

Introduction

Since the early 1990s, the World Trade Organization (WTO) agreement coupled with environmental initiatives have imposed the need for reform in crop farming. The aim of this reform should be to produce internationally competitive crops and deal with ecological disturbances caused by agricultural activities. In addition, the «Rural Europe: Perspectives for the Future» convention of Cork, Ireland, in November 1996, with delegates from all European Union (EU) member countries, the rest of Europe, Canada, USA, and Japan, concluded with a declaration that future agricultural policies must be more closely oriented to the goals of environmental protection and rural development (Anonymous, 1997a).

Thus, there are pressures for agriculture to become both competitive and sustainable, environment-friendly and a vehicle for rural development. The extreme case of environment-friendly agriculture is organic or biological, totally excluding the use of chemical inputs in production. However, the path to sustainable agriculture is a matter that needs careful consideration for it is possible, under the mask of environmental protection, to deteriorate the farming system and crop productivity in a way which may exaggerate farm problems and human starvation (Scientists' Review 1990; Galanopoulou-Sendouca, 1995a).

The European Commission is funding research projects for sustainable agriculture and it also applies mechanisms and directives towards free market based agricultural production, in line with its new Common Agricultural Policy (CAP) goals. Simultaneously, the Commission is concerned with issues of development in the rural European countryside.

Under these international developments, it is interesting to look specifically at crop production and the efforts undertaken to formulate approaches to achieve higher competitiveness and sustainability of cotton cultivation. It must also be noted that cotton cultivation involves a highly erosive chemical process.

Cotton (*Gossypium hirsutum* L.), although of a tropical or subtropical origin, is cultivated mainly in temperate regions. Today, the world cotton acreage exceeds 30 million hectares, with leader countries the USA, China, India, and Pakistan, which produce approximately two thirds of the total production. Cotton plants produce natural fibers of incomparable properties so that they cover roughly fifty percent of the global fiber use. They also produce cotton seed which is a rich source of oil and protein for animal and human consumption. Cotton is considered a stock product and its production affects the economic development and prosperity of many countries (Galanopoulou-Sendouca, 1992).

For Greece, cotton is the top agro-industrial product of great importance for both agriculture and the national economy. It is cultivated in an area that exceeds 0,4 million hectares, giving basic employment and satisfactory agricultural income to 80,000 - 100,000 rural families. Approximately 150,000 urban workers are employed in primary and secondary cotton production and use. Cotton contributes substantially to the Greek textile industry, and is an important source of foreign exchange (Kalogiros, 1994). It must be noted that Greece, although a geographically marginal area for cotton cultivation, is among the world leader countries in terms of both yield and quality of Upland cotton. Moreover, it is the main cotton producer member state of the European Union (EU), contributing more than 80% of total production.

The main factors for cotton growth in Greece have been: the foundation of the Hellenic Cotton Board and the Cotton Research Institute, with the successful role these institutions have played in the improvement of suitable cotton varieties and cultivation techniques; the development of the local textile industry; the complete mechanization of production; and, Greece's entry into the EU which allowed it to enjoy high producer prices due to the application of subsidies (Galanopoulou-Sendouca, 1994).

However, the intensification of cotton cultivation which aims mainly at yield maximization, has led to an unreasonable chemical fertilization and over-irrigation of crops, the overuse of chemicals for plant protection, the employment of heavy farm machinery causing soil compaction, and the elimination of crop rotation with the subsequent extinction of many plant varieties and species.[1] The environmental consequences of intensive farming have become obvious since the early 1980s, and are increasingly important to policy makers AND farmers, as well as the general public.

Cotton has emerged as the major consumer of agrichemicals not only for yield maximization purposes, but also because it has a comparatively longer growing period compared to many other crops. Its longer stay in the field and its specific fruiting behavior increase its vulnerability to insects pests and diseases. In addition, as the biological balance gets disturbed, pests develop resistance to chemicals, the insect pattern changes, etc, giving rise to multifarious production problems. Consequently, in most countries, agrichemical costs have increased dramatically, constituting about 50% of the total costs of seed cotton production - excluding land rents and other fixed costs (ICAC, 1994).

The costs of production due to high inputs, have increased to the extent that cotton farms are loosing their profitability against other farms. Besides, with the reformed cotton status since 1995, according to the new CAP, farmer cotton prices in Greece are decreasing as much as the total seed cotton production exceeds the level of 787 kilotons (257 for Spain) which is very low in comparison to the actual cotton production of 1,200 kt. Taking into account the fact that Greece is among the highest yield cotton producing countries worldwide (~ 3 t/ha seed cotton) and further yield increases are not easily feasible, it is even more important to reduce production costs.

In order to reduce production costs and avoid any impacts which agrichemicals may have on the environment, cotton growing with low inputs is of high priority. Simultaneously, in a world of liberalized agricultural trade, lower production costs associated with low input use and high cotton yields increase the competitiveness of cotton.

This chapter suggests, among other things, that potential reductions to the level of inputs used in cotton production in Greece and other Southern European countries could be made without sacrificing current high yields. Simultaneously, however, the option for organic cotton farming implies higher production costs and, thus, the need to enjoy higher market prices. Under these circumstances, while maintaining rural incomes at satisfactory levels and ameliorating ecological disturbances, cotton may continue to be an internationally competitive crop.

The remaining parts of this chapter present cotton growing systems with low inputs which may have a favorable effect on production costs directly, alternative approaches to cotton growing which affect production costs indirectly, and, lastly, the conclusions with policy recommendations.

Cotton Growing Systems with Low Inputs

Cotton growing systems with low inputs include low fertilization and irrigation, drip irrigation, crop rotation, minimal tillage, and the creation of cotton varieties resistant to biotic and abiotic stresses. Each of these are examined below, highlighting evidence from Greece and elsewhere using experimental data.

Low fertilization and irrigation

Cultural (cultivation) management for cotton and other crops has been greatly enhanced in the last 15 years by the use of dynamic crop growth and development simulation models in the soil-plant-atmosphere continuum. GOSSYM/COMAX is a combination of a crop simulation model (GOSSYM) and an expert system (COMAX) which manages the production system based on the simulation provided by GOSSYM. The model was created using research and data bases with cultivars grown in the Mississippi Delta area but was also calibrated and validated for other crop growing areas of USA (Gertsis, Galanopoulou-Sendouca, and Papathanasiou, 1994). The model was evaluated for a period of 3 years (1994-96) under the soil and climatic conditions of Greece and Spain, the main cotton producing countries in the EU.[2] Field experiments were conducted in three Greek and three Spanish locations representing the major cotton production areas in both countries. The experimental variables comprised three cotton varieties growing in two drip irrigation levels (for Thessaly: ~ 900 and 1.800 m^3/ha), two fertilization levels (for Thessaly: 140-70-70 and 240-120- 120 units NPK/ha), and two plant population levels assigned in a split-split-split plot experimental design.

Although calibration and validation of the model has not yet been completed, the 3-year experimental results provide substantial information and support the hypothesis that low input production systems are feasible for optimal cotton production. Conventional levels of major production inputs, such as nitrogen fertilizer, and irrigation water did not result in yield increases as compared to lower levels, and in most cases low input treatments led to an increase of yield in the first pick which represents the most important part of the total cotton yield and the best quality of cotton fiber (CPR, 1996). Table 8.1 presents the results of seed cotton yield in 1996 in three Greek locations as affected by the two main treatments: farmer, representing a management schedule used by the typical farmer, and GOSSYM, representing a management decision scheme based on the low input rationale supported by GOSSYM (Gertsis *et al*, 1997).

TABLE 8. 1: Seed cotton yield in three locations (kg/ha)

	LOCATION								
	Velestino			*Palamas*			*TEI-Sindos*		
Pick	Farmer	COSSYM	*Sig*	Farmer	COSSYM	*Sig*	Farmer	COSSYM	*Sig*
1st	1440	1710	*ns*	1630	1920	*ns*	1330	1830	*
2nd	590	400	**	930	900	*ns*	120	130	*ns*
3rd	150	80	*	290	140	*	530	500	*ns*
Total	2190	2180	*ns*	2850	2950	*ns*	1980	2450	*

Sig= Significance level
ns= Not significant
* = Significant difference for $p= 0.05$
**= Significant difference for $p= 0.01$

In addition to the above, a 5-year pilot project for the reduction of nitrogen pollution from cotton production is being currently applied in the region of *Thessaly* since 1995, in the framework of 2078/92 EU Regulation, concerning cotton cultivation friendly to the environment. The project's main objective is the reduction of surface and underground water nitrate pollution, due to excessive nitrogen fertilization in cotton cultivation. Necessary prerequisites for the application by the farmers are: adoption of crop rotation so that cotton cultivation should not exceed 3 successive years; drip or sprinkler irrigation so that the application of fertilizers could be combined with irrigation; and, reduced nitrogen fertilization using 103 kg/ha, 50 of which should be applied as basic fertilization during cotton sowing, and the rest dissolved in water during irrigation in 5 doses from the 7th pair leaf stage up to the boll maturation stage. Preliminary results show that reduced nitrogen fertilization leads to yield decrease when applied with conventional methods. However, when it was applied as above with advanced technology and at the appropriate phaenological stages the yield was up to 15% higher than the conventional high level nitrogen fertilization.[3]

The CAP provisions for lower nitrogen fertilization are not irrelevant with some decisions as free prices of fertilizers, land set-aside subsidized systems, and acreage instead of product subsidies. However, it must be noted that adoption of cotton acreage subsidies, instead of the existing system of production subsidies, should result in cotton quality deterioration and consequently reduced cotton competitiveness.

The problem of water pollution with nitrates is intensified by excessive irrigation, which not only has adverse effects on crop productivity, but also leads to dangerous shortage of the precious water reserves. Agriculture policymakers in many developed countries are close to imposing fees on farmers who exceed the economic and environmental optimum irrigation level.[4]

Drip irrigation

Recently, drip irrigation is expanding rapidly in Greece especially in Thessaly where it covers approximately 50% of the cotton cultivation. It is mainly used in regions with intense water shortage problems and lack of irrigation networks but with high crop yielding capacity, so that covering the expenses of buying the system can be achieved during its relatively short life span.

In places such as Israel, where the shortage of water is very serious and where the drip irrigation system has its origin, the percentage of the area irrigated by drip irrigation is roughly 60-70%. Compared with sprinkler irrigation, yields of crops irrigated by drip irrigation are generally 15-20% higher, according to research work carried out for at least 15 years (Goren, 1994). The main advantages of drip irrigation are: effective use of water (approximately 40% less water in comparison to surface systems as there is not much waste due to evaporation or its movement below the root system), and effective fertilization and weed control through the system (leading to lower cost of input applications).

The continuous cultivation of the same crop deteriorates the soil and makes it poorer due to its specific needs of natural sources, especially soil nutrients. However, in the framework of intensive agriculture, crop rotation has been replaced in most cases by monoculture of 'competitive' subsidized crops. This fact is also due to the false view that crop profitability is calculated on an annual basis instead of a period of certain years.

The yield decrease in monoculture systems is attributed to «soil sickness». Plant growth and development is prohibited as plant resistance to diseases and insects is decreased. «Soil sickness» is also connected with the fact that different crops have different nutrient demands. Another dimension of «soil sickness» is the root excretions of toxic substances from certain plants with *allelopathic* adverse effect on other plants' growth. Allelopathia has greater effect on weeds control (Lampkin, 1992).

The most important advantages of crop rotation are:

a) Conservation or even improvement of soil fertility.

b) Control of plant pests, diseases and weeds due to great differences among crops. Many pests and diseases infest only certain crops, so that crop rotation is a main factor of integrated pest management. Crop rotation is more effective in disease control than in insects. For example, severe soil infestation by the fungus *Verticillium dahliae Kleb,* resulting to great cotton yield losses, obligates cotton farmers to change their crop for certain years.

c) Yield and quality increase of crops. Adding legumes, e.g., as green manure or forage in different cropping systems, can fully open up the limited land resources, increase the potential of soil and improve soil fertility. In China, legumes used as green manure incorporated into the soil 22,5 t/ha biomass and decreased N fertilization for following cotton crop by 50%. Also, broad beans (*Vicia faba L.*) or peas (*Pisum sativum L.*) are interplanted with wheat in the autumn or early winter and picked as fresh pods for vegetables. The residues are incorporated as basal manure for the succeeding cotton crop the next spring. Experiments showed that inter-planting broad bean gained 5.4 t fresh pods/ha, a bonus yield, and 16.8 t fresh matter/ha which provided additional organic matter and nutrients as shown in Table 8.2 (Chen, 1992).

TABLE 8. 2: Bonus yields of broad bean interplanting

Treatments	Broad beans: Pods	Broad beans: Biomass	Unginned cotton	Input cost	Gross return	Net return
With broad bean	5.4	16.8	2.45	361.1	1,559.7	1,198.6
Without broad bean	—	—	1.75	277.8	972.2	694.4

Source: (Chen, 1992). Quantities are in t/ha while costs and returns are in $/ha.

In Greece cotton cultivation has become, due to high prices, a monoculture in some regions, e.g. Thessaly, and this situation has led to «soil sickness». Even wheat that was used in rotation with cotton has now disappeared. Today, legumes and cover crops must be included in rotation systems to minimize inputs in cotton cultivation. An appropriate crop rotation system may thus lead to the increase of cotton farmer income in the whole period of rotation due to lower cost and to higher cotton prices as a result of reduced cotton production.

Minimal tillage

The intensive use of unsuitable machinery in connection with cotton monoculture initiated problems of soil compaction, which led to adverse effect on plant rooting and consequent crop yield.

Farmers in Central Greece cultivate their fields with irrigated cotton and use wheat as a break crop every 3-4 years. In recent years, delay of cotton harvesting, due to the use of late maturing varieties and cultivation techniques, pushed farmers to speed wheat sowing without the conventional tillage especially in wet years. A method of establishing wheat after cotton by broadcasting the seed without any previous stalk chopping and tillage and seed incorporation by a light cultivator in comparison with conventional tillage, was studied for three years (1993-95) at the University of Thessaly in Central Greece. The results showed that yields in plots with unchopped stalks did not differ significantly from those in plough plots. The no-tillage methods presented a lot of advantages in terms of timeliness, low cost and energy consumption, and appear to be a good alternative practice allowing farmers to increase their margins and adapt to the new conditions of the world market after the new CAP and GATT agreements (Gemtos *et al*, forthcoming).

Cotton varieties resistant to biotic and abiotic stresses

Breeding and use of improved varieties, adapted to the demands of cultivation and consumption, is the most effective and inexpensive factor for farmer cost reduction and for increase of crop productivity. Breeding of cotton varieties is of additional importance in Greece, as foreign varieties bred under more favorable ecological conditions have only a limited possibility of being adapted to the specific and generally unfavorable Greek climatic conditions for cotton growth and development. This option is verified by the fact that from 1930, when cotton started being cultivated systematically, until recently, Greek varieties dominated cotton cultivation, which is an exception among other crops. The Greek varieties were improved and adapted to local conditions and have helped Greece enter the world's top yield producer list with an excellent fiber quality of Upland cotton. After freeing cotton variety trading, many foreign varieties started being introduced in Greece every year. Although the use of most of them does not exceed 1-2 years, they increase their total acreage percentage due to better marketing and seed quality in comparison with local varieties. Foreign varieties represented only 10% of the total acreage in 1990 and reached approximately 60% in 1996. The use of multiple varieties destroyed the uniformity of Greek cotton and deteriorated its quality (Galanopoulou-Sendouca, 1994).

Until recently, plant breeders have been trying to create varieties of high response to synthetic fertilizers and generally to high inputs. Research is now oriented to the creation of low nutrients demanding varieties which are also generally resistant to pests, diseases, weeds, and abiotic stresses.

Resistance to insects

Promising results come from the development of transgenic cotton varieties containing Bt genes (from *Bacillus thuringiensis*) for controlling of Lepidopteran pests in general and particularly *Heliothis* spp which is largely responsible for high yield losses. The insecticidal activity of Bt gene comes from a toxic protein very specific to lepidopteran insects. The Bt gene has been confirmed to be non - toxic to useful insects and does not carry any undesirable environmental effects according to the Monsanto Company. Trials conducted by Monsanto on transgenic cotton have shown that transgenic cotton tolerated lepidopteran insects as well as non-transgenic cotton sprayed every week (Buehler, 1993). Two cotton varieties created by Monsanto Co. were cultivated in USA in 1996/97 in an area of 700 thousand hectares equivalent to approximately 12% of the total area. Such varieties started being cultivated also in other countries and they are now ready to enter Greece. Creation and testing of cotton varieties with Bt genes is a research objective in many countries nowadays (ICAC, 1997). However, some reports show that the cotton bollworm (*Heliothis armigera*) has developed resistance to Bt and that limitation of bollworm population results to the population increase of other insects due to biological balance (ICAC, 1993).

Cotton drought tolerance

The continuous limitation of precious water sources increases the importance of drought resistance. Researchers in U.S.A have discovered a technique that may help to identify strains of cotton that can survive drought and produce more and better lint (Anonymous, 1992). Moreover, promising information comes from Israel for newly bred cotton varieties tolerant to drought. Finally, resistance genes to dry conditions have been discovered in wild cotton species as G. *anomalum* and *G. thurberi* (Musaev, 1993).

Alternative Approaches to Higher Input Productivities

Alternative approaches to cotton cultivation which affect production costs indirectly include: early cotton sowing under plastic film, cotton growing on narrow-row spacing, and cultivation of organic and colored cotton.

Early cotton sowing under plastic film

This method is broadly used in Spain to guarantee seed germination under early sowing and expanded growing season (Marquez, 1994). For Greece, especially its northern regions which are characterized by a small growing season, this system contributes to early and uniform plant emergence, an important component for the crop's success. Research data from the 1970s (Galanopoulou-Sendouca, Markousis, and Chlichlias, 1978) as well as evidence from farmers who have adopted the recently mechanized method, supports the use of plastic film in the early stage of cotton cultivation. However, there are same contradictory results, and further investigation is needed. Recent research at the University of Thessaly showed that higher yield overcompensates the cost of plastic use, and that earlier cotton maturation guarantees successful cotton harvesting (Stathakos and Galanopoulou-Sendouca, 1996). However, the continuous use of plastic films will create environmental problems which may be solved by the production and use of either biodegradable plastic or films from starch, e.g. cornstarch (Anonymous, 1991).

Cotton growing on narrow-row spacing

The conventional cotton growing system with rows 1m apart has been imposed, mainly by the adoption of cotton pickers adapted to this row-distance. Many experiments in Greece and elsewhere provide evidence that narrow-row systems are superior because they guarantee better plant distribution and earlier canopy closure in comparison to the conventional system (Galanopoulou-Sendouca *et al*, 1980). The superiority of narrow-rows is more evident when the prevailing conditions prohibit the canopy closure by mid-July in 1m rows. Such cases are expected to become more frequent with the use of low nutrients demanding short growth cotton varieties and the use of low inputs growing systems which will reduce plant growth.

Since the 1980s in California, there has been an increasing trend in the use of transformed cotton pickers which can successfully harvest cotton sown in 0.76cm rows. (Weir, 1991). Research data showed a yield increase in rows 76 cm apart compared to rows 102 cm apart. This superiority of better plant exploitation of soil and solar energy is due to the spatial uniform distribution of plants. (Heitholt, Pettigrew, and Meredith, 1993). Early maturity was also confirmed and attributed to limited plant growth and to the lower first fruiting branch (Heitholt, Pettigrew, and Meredith, 1993; Robinson, 1991, Williford, 1992). Since 1997 this system is also being evaluated in Greece in the framework of a research project at the University of Thessaly.

Cultivation of organic and colored cotton

Organized efforts to grow cotton without agrichemicals began around 1990. The ICAC (International Cotton Advisory Committee) began reporting on organic production in early 1993, and in October 1996 it published in the *ICAC Recorder* a compilation of articles of its Technical Information Section. In 1996 more than 8,000 hectares of organic cotton were grown the USA, the largest cotton producer in the world. Organic cotton is also produced in other countries such as Australia, Egypt, India, Turkey and recently in Greece. Certification of production and processing of organic cotton is handled mainly by private organizations. Currently, organic cotton is knitted into T-shirts, sweaters, infant wear, towels etc. According to research data and evidence up to 1994, maintaining yields in organic cotton production is not easy. The average yield decrease is estimated at 25%. Also, in most cases the cost of production is at least 10% higher, mainly due to manual operations, especially for weed control and expensive non-conventional means of insect control. It appears that it will not be economical to produce organic cotton unless it gets a 43% higher price over conventional produce. So far, high prices and subsidies compensate yield loss and additional production costs in most cases. Organic production demands a higher level of skill than conventional production, but no systematic research has appeared as yet to provide guidelines for shifting from conventional to organic production. The success of organic production rely, mainly, on the use of fertile soils and of suitable varieties, and on the absence of pests, diseases, and weeds. Similarly, other components of integrated pest management could contribute to the success of organic production.

Colored cotton grown under organic conditions is also available on the market for spinning and is more desirable for organic processing because finishing and dying are mostly chemical processes. Presently, only brown in various shades and green colors are available on a commercial scale. Blue and yellow cotton is expected to be available in the next few years. Breeding objectives in colored cotton include improving color range fiber quality and yield potential. Colored cotton is grown in many countries but with aggressive campaigning by Israel. Total area under organically grown colored cotton was estimated to be between 2,500 - 2,800 ha during 1994/95 (ICAC, 1996).

Organic cotton started being produced in Greece from 1993 on a small scale (Galanopoulou-Sendouca, 1995b; ICAC, 1996). Seed cotton yield was ranged between 1,500-3,500 t/ha and severe problems arose mainly from weeds which were controlled manually, thus the cost of production was increased. Colored cotton is also being evaluated under experimental conditions. It should be noted that cotton production in Greece is more environmentally friendly as regards dealing with insect problems; normally only one spraying per year is sufficient versus more than eight sprayings realized in other countries.

Conclusions and Policy Options

Experimental and other data presented in this chapter indicate that savings in water, seeds, and fertilizers costs could be made without necessarily sacrificing currently high cotton yields. This translates into higher input productivities and a more competitive and sustainable production, which recent international agreements dictate. For this to be accomplished, however, cotton farming would need to adopt growing systems with low inputs, such as low fertilization, drip irrigation, etc.

Alternative production techniques, such as cotton sowing under plastic film, and generally enhancing the quality of the product could help indirectly toward the same goals. Organic cotton growing constitutes an exception as regards competitiveness, for it entails higher production costs. Only higher market prices, usually enjoyed by organically produced crops, can reverse the trend.

Under these conclusions, policy options should be carefully designed. Undoubtedly, the aim of yield maximization of cotton production in Greece should be abandoned. Instead, growing systems aimed at optimal combinations of production costs and yields must be adopted by cotton growers. Cotton cultivation must become also more friendly to the environment with low input use. Total seed cotton production should not exceed 1 million tons per year in order to avoid drop in prices.

Cotton cultivation should be expanded in regions with high adaptation where satisfactory yields can be achieved with low cost. Conversely, expansion of cotton cultivation in areas of poor adaptation, as happened recently, leaves cotton growers in these areas without profitability and it also deteriorates the conditions for other cotton growers, as it leads to increases in the total volume of production and subsequent drops in market prices.

The quality of Greek cotton could recover its recent levels and business risks may be reduced, by using varieties with uniform fiber properties, and by differentiating cotton farmer prices according to the quality of seed cotton so that farmers should be stimulated to select varieties of high quality and to offer to the gins good quality of seed cotton. Delaying cotton maturity, by the use of late maturing varieties and cultivation techniques, increases theoretical productivity, but very often leads to yield and quality losses. It also prohibits cotton cultivation from being successfully included in crop rotation systems. Improvement of the seed propagation system is absolutely necessary, so that Greek varieties with superior fiber quality can compete successfully with foreign varieties which have inferior fiber quality, but better quality of sowing cotton seed. Breeding of suitable new cotton varieties should be enhanced to meet the changing demands of cotton farming and consumption.

Finally, economic evaluation of alternative farming systems under Greek conditions should be intensified before they are broadly adopted by cotton growers.

Notes

[1] Until 1962, on an average basis, the nitrogen uptake by the crops was more or less equal to the amount of fertilization, while in 1987 the amount of nitrogen input was twice the amount of output (Gatzianas, 1993).

[2] Project No AIR3 -CT93-0936. The project was funded by the EU and is being coordinated by the Laboratory of Agronomy, University of *Thessaly*. The objective of this project is the creation of a European version of GOSSYM and the application of the model by the cotton producers in Greece and Spain for the optimization of cotton production in the context of sustainable agriculture and input reduction in the production system.

[3] Personal communication with Dr. K. Karatzos, project coordinator.

[4] In Greece, recently (25 June 1997) a Decision signed by six ministers was issued in the official Governmental Gazette, according to which necessary terms are described, and measures will be taken to reduce water pollution and to protect water from further pollution caused from nitrates of agricultural origin (Anonymous, 1997b). The decision is based on EU Directive 91/676. The fulfillment of the objectives will rely on the use of agricultural management measures such as: appropriate land use, including crop rotation systems, balance between perennial and annual plants, cover crops during rainy periods (for nitrogen absorption by the plants instead of water pollution as well as for soil erosion), stimulation of the farmers for reasonable fertilization designs and for keeping accounts on fertilization data. Another measure of high priority will be the limitation of the fertilization amounts permitted for each crop according proper agricultural practice and the prevailing ecological conditions of the particular region.

References

Anonymous, 1991. «In Plastic Bags : Cornstarch,» *Farm and House Research*, Agricultural Experiment Station, S. Dakota State University, 41(2) : 5-12.

Anonymous, 1997a. «Special File for the Cork Convention,» *Geotechnical Information*. 100: 12-16 (in Greek).

Anonymous, 1997b. «Measures and Terms of Water Protection from Nitrate Pollution of Agricultural Origin,» *Geotechnical Information*, 100: 17 - 22 (in Greek).

Buehler, R. E. 1993. «The Development of Transgenic Cotton Containing a Bt gene for Control of Lepidopteran Pests,» *The ICAC Recorder*, 11(2): 3 - 4.

Chen, L. 1992. «Crop Sequences for Sustaining Soil Resources in China,» *Proceedings of the International Crop Science Congress*, Iowa State University, Ames, Iowa, pp. 101 - 102.

CPR (Consolidated Progress Report), 1996. «Validation and Application of GOSSYM - A Cotton Growth Simulation Model to Optimize Cotton Production Under the Soil, Climatic, and Cultural Conditions of Greece and Spain,» *European Commission*, DG XII, Project No AIR3 -CT93 - 0936

Galanopoulou-Sendouca, S., S. Markousis, and A. Chlichlias, 1978. «Effect of Sowing Date and Row Covering with Polyethylene Film in Cottton (*Cossypium hirsutum L*),» *Agricultural Research*, 2 (2) : 167 - 180 (in Greek).

Galanopoulou-Sendouca,S., A. G. Sficas, N. A. Fotiadis, A. A. Gatzianas, and P. A. Gerakis, 1980. «Effect of Population Density, Planting Date, and

Genotype on Plant Growth and Development of Cotton,» *Agronomy Journal*, 72 : 347 - 353.

Galanopoulou-Sendouca S. 1992. *Specific Agronomy II.* Faculty of Agriculture, University of Thessaly (in Greek).

Galanopoulou-Sendouca, S. 1994. «Cotton Varieties and New Cultural Techniques to Increase Greek Cotton Competitiveness,» Proceedings of the Congress *Greek Cotton in Europe*, Geotechnical Chamber of Greece, Thessaloniki, pp. 39 - 56 (in Greek).

Galanopoulou-Sendouca, S. 1995a. «Intensive, Alternative, and Ecological Agriculture,» *Vima*,The Agronomists' Union of Larissa Gazette, 17 : 1 - 5 (in Greek).

Galanopoulou-Sendouca, S. 1995b. «Ecological Cotton Cultivation,» Proceedings of the Scientific Symposium *Days of Cotton, 1994*, Prefecture of Karditsa, pp. 69 - 78 (in Greek).

Gatzianas, A. A. 1993. «The Need to Reduce Nitrogen Fertilizer for Good *Agricultural Practice in Greece,»* in V. Jordan (ed*) Agriculture: Scientific Basis for Codes of Cood Agricultural Practice,* CEC. EUR 14957EN, pp. 45 - 49.

Gemtos, T. A., S. Galanopoulou-Sendouca, and C. Kavalaris, (forthcoming). «Wheat Establishment after Cotton with Minimal Tillage,» *European Agronomy.*

Gertsis, A., S. Galanopoulou-Sendouca, and G. Papathanasiou, 1994. «A Cotton Crop Simulation Model for Cultivar Evaluation and Cultural Management,» proceedings of 5th congress of Greek Scientific Society for Genetics and Plant Breeding «Contribution to the environment deterioration and human starvation», Volos, pp. 248 - 255, (in Greek).

Gertsis, A. S. Galanopoulou-Sendouca, G. Papathanasiou, and A. Symeonakis, 1997. «Use of GOSSYM - A Cotton Growth Simulation Model to Manage a low Input Cotton Production System in Greece,» Proceedings of the First European Conference for Information Technology in Agriculture. (Posters Session). *The Royal Veterinary and Agricultural University Copenhagen, Denmark* 15-18 June 1997, pp. 359 - 362.

Goren, M. 1994. «Drip irrigation,» Proceedings of congress «Greek Cotton in Europe» Geotechnical Chamber of Greece, Thessaloniki, pp. 97 - 100.

Heitholt, J. J., W. T. Pettigrew, and W. R. Meredith, 1993. «Growth, Boll Opening Rate and Fiber Properties of Narrow - row Cotton,» *Agronomy Journal,* 85: 590 - 594.

ICAC (International Cotton Advisory Committee), 1992. «Short Notes,» *The ICAC Recorder*, 10(3): 3- 13.

ICAC (International Cotton Advisory Committee), 1993. «Research Priorities in Australia,» *The ICAC Recorder*, 11(3) : 2 - 5.

ICAC (International Cotton Advisory Committee), 1994. «Organic Cotton Production II,» *The ICAC Recorder*, 12(2): 5-9.

ICAC (International Cotton Advisory Committee), 1996. «Growing Organic Cotton,» *Report of ICAC,* Washington, D.C., October.

ICAC (International Cotton Advisory Committee), 1997. «Current Research Projects in Cotton,» *Report of ICAC*, Washington, D.C., October.

Kalogiros, K. 1994. «Importance of Cotton Cultivation to the Greek and World Economy,» Proceeding of the congress *Greek Cotton in Europe*, Geotechnical Chamber of Greece, Thessaloniki, pp. 13 - 23 (in Greek).

Lampkin, N. 1992. *Organic Farming*, Ipswich, UK: Farming Press Books.

Marquez, F. P. 1994. «Growing Cotton Under Plastic Film,» Proceedings of the congress *Greek Cotton in Europe*, Geotechnical Chamber of Greece , Thessaloniki , pp. 67 - 76 (in Greek).

Musaev, D. A. 1993. «Genetic and Breeding Techniques to Improve Cotton Production under Irrigated Conditions in Usbekistan,» *52nd Plenary Meeting of the International Cotton Advisory Committee*, New Delhi, India, pp. 5-9.

Robinson, E. 1991. «The Development of the First Narrow-Row Cotton Picker Involves More Than Just Row Spacing,» *Cotton Grower*, July 1991, pp. 12 - 13.

Scientists' Review. 1990. «Alternative Agriculture,» *Council of Agricultural Science and Technology* , Washington, D.C., Special publication No 16.

Stathakos, T. and S. Galanopoulou-Sendouca, 1996. «Row Covering with Polyethylene Film for Early Cotton Sowing,» *Agricultural Technology*, special issue on «Cotton - Maize' 96» pp. 60 - 67 (in Greek).

Weir, B. L. 1991. «Narrow-Row Means Higher Yields,» *Cotton Farming*, Special Issue, 96: 8 - 9.

Williford, J. 1992. «Production of Cotton on Narrow - row Spacing,» *Transaction of the ASAE*, 35 (4): 1109 - 1111.

Part II
Fisheries and Forests

Chapter 9

Stochastic Trends in Productivity Analysis: An Application to the Fleet of Bottom Trawlers

Panos Fousekis and *Christos Pantzios*

Introduction

The use of fishing nets to sweep the sea bottom is a fishing method widely criticized for the adverse environmental impact it may cause via indiscriminate harvesting or overharvesting of fish populations. In Greece, although the fleet of bottom trawlers constitutes less than 5 percent of the total number of vessels operating in the country, it accounts for almost 20 percent of the total catch. The activities of the fleet are closely monitored by National and EU authorities because of their serious impact on fish populations. A number of additional restrictions have also been imposed on this fishing method via a set of EU regulations since the early 1990s.

The relationship between fishing effort and harvest is central in to the economics and management of fisheries (Clark, 1985; Anderson, 1985; Squires, 1987). This relationship, however, is inherently stochastic. Technological innovations as well as permanent and temporary changes in the level and the density of the fish stocks make it impossible to determine *a priori* the level of harvest resulting from a given fishing effort (Pindyck, 1984; Conrad and Clark, 1991, Bernard and Jones, 1996). Therefore, standard econometric approaches which rely on *deterministic trends* to capture the effects of technological improvements and fish density changes in the harvest-effort relationship may lead to biased and inconsistent estimates and forecasts (Clark and Youngblood, 1993; Hamilton, 1996).

This chapter has two objectives. Primarily, it aims at investigating the relationship between fishing effort and harvest associated with the

Greek bottom trawlers during the period 1964-95 via an empirical analysis of the determinants of the overall productivity of this fishing activity. At the same time, it attempts (while largely remaining within the confines of a conventional economic approach) to introduce resource preservation into the framework of analysis. Although such considerations enter the analysis only indirectly, our approach does account for reservation concerns, traditionally absent from economic modeling for key measures such as overall sectoral productivity, which is a measure of competitiveness.

In order to enhance the flexibility of modeling the fishing effort-harvest relationship of bottom trawlers so as to account for the presence of environmental factors, we introduce the concept of a *stochastic trend*. Signal extraction techniques are used to estimate the relationship in question and to decompose the overall productivity of the bottom trawlers' fishing effort into a number of effects, namely, a scale effect, a technology and fish stock density effect, and a temporary effect.

The chapter is organized as follows. First, it presents the concept of a stochastic trend and develops a variant of the standard harvest-effort model that is used in the empirical analysis. Following that is a brief description of the fleet of bottom trawlers in Greece that describes the evolution of variables of interest such as harvest, effort, and productivity. Next, the Kalman Filter Algorithm is used to estimate the harvest-effort relationship and to decompose the productivity of fishing effort during 1964-95. Finally, conclusions and policy implications are drawn.

The Theoretical Model

The theoretical model employed is based on the following assumptions:

(a) The harvest-effort relationship is given as,

$$(1) \quad Q_t = E_t^r \, y_t$$

where Q is harvest, E is the level of fishing effort, r is the elasticity of catch with respect to effort, t is time, and y stochastic term which represents all factors affecting the relationship between catch and effort, besides the amount of effort.

(b) The factors represented by y can be classified into two categories. The first category includes factors causing permanent changes in the productivity of effort. These factors include technological progress and permanent changes in the density and the level of fish stocks. The second category includes factors leading to temporary changes in the productivity of effort. These factors include weather conditions as well as temporary changes in the density and the level of fish stocks. The assumption that y represents two types of factors can be formalized by expressing y as

$$(2)\quad y_t = g_t\, e_t$$

where g is the permanent component of y and e is the transient component.[1]

(c) The permanent component, g, evolves as a random walk with drift, that is,

$$(3)\quad g_t = g_{t-1} + b_t + n_t$$

where b is the drift and n a random shock.

(d) The term b is stochastic and evolves as a random walk without drift, that is,

$$(4)\quad b_t = b_{t-1} + z_t$$

where z is a random shock (e). The stochastic terms e, n, and z are uncorrelated with each other and normally distributed with zero means and variances S_e, S_n, S_z respectively.

Equations (2), (3), and (4) constitute a system of difference equations with solution

$$(5)\quad y_t = [\sum_{i=1}^{t} n_i + t z_1 + (t-1) z_2 + \dots\dots\dots z_t] e_t .$$

A comparison between equation (2) and equation (5) reveals that the permanent component, g, is just the term in the brackets on the right hand side of (5). Given that n and z are stochastic shocks, the permanent component is also stochastic and it is called *stochastic trend*. Like the deterministic trends, g is characterized by its *level* and its *slope*. The slope is the logarithmic derivative of g with respect to t and gives the time rate of change of g. When the variances of z and n

are zero, the stochastic trend collapses to the standard deterministic trend.

Having defined the stochastic trend, we return now to the decomposition of the fishing effort productivity. The rate of growth in the productivity of effort is

$$(6)\quad \hat{P} = \hat{Q} - \hat{E}$$

where $\wedge$ denotes the time rate of change and $P = Q/E$. Given the functional forms of equations (1) and (2) the decomposition of fishing effort productivity becomes

$$(7)\quad \hat{P} = (r-1)\hat{E} + \hat{g} + \hat{e}.$$

The first term on the right hand side of (7) is the scale effect while the third one is the temporary effect. We call the second term on the right hand side of (7) 'productivity effect.' [2]

The Greek Bottom Trawler Fleet

According to official data published by the National Statistical Service of Greece (NSSG), the bottom trawler fleet in the country currently consists of 375 vessels. The total Horse Power (HP) of these vessels is 146140, the total Gross Registered Tonnage (GRT) is 23656, while total employment is 1396.

The quantity of fish harvested by the fleet of bottom trawlers has more than doubled since 1964, the year that statistical monitoring of the fleet's operation began in Greece. The average annual rate of growth in harvesting during the period 1964-95 was 2.77 percent, as seen in Figure 9.1.

Employment exhibits substantial variability over the years. From 1964 until the mid-1980s there was a clear downward trend in employment. That trend, however, has been reversed in recent years. The average annual rate of change in employment, during the period 1964-95, was -0.3 percent. The number of vessels in the fleet exhibits substantial variability as well. Over the last ten years, however, there has been a positive trend in the number of vessels. The annual average rate of growth in the number of vessels, during the period 1964-95, was 0.1 percent. The HP has increased more than three times since 1964. The average rate of growth in HP was 3.74 percent per annum over the examined period (Figure 9.2).

Figure 9. 1: Harvests of the Bottom Trawlers in Greece (in Metric Tons)

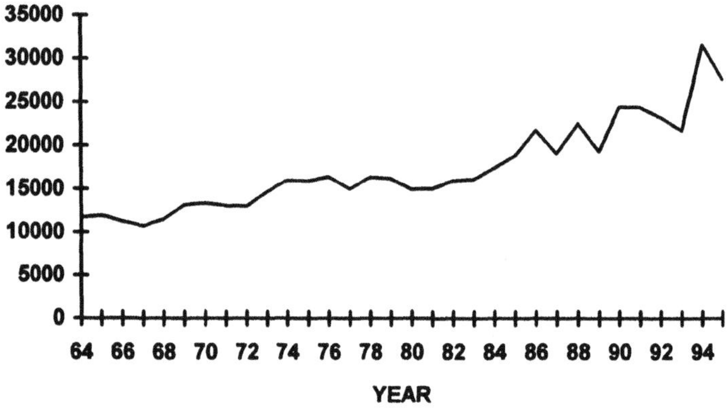

Figure 9. 2: Horse Power (HP) of the Bottom Trawel Fleet in Greece

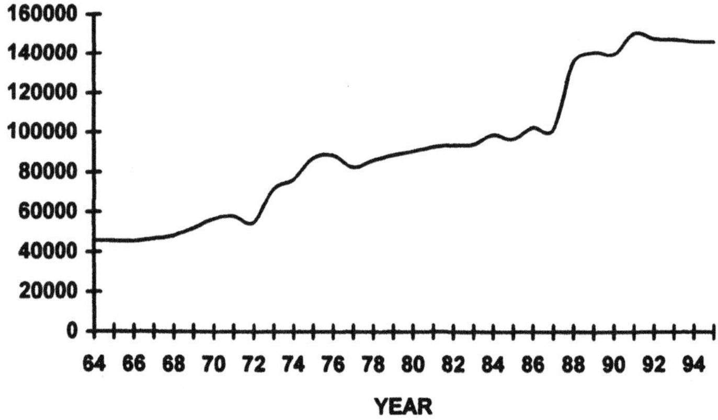

HP plays a very important role in the vessels' capacity to harvest fish since it is used to draw the special nets of trawlers along the sea bottom. Thus, there is a direct relationship between the level of HP and the quantity of fish harvested. Because of this relationship, the amount of fishing effort exercised by trawlers is often identified with the level of HP. The EU in its attempt to preserve the fish stocks has recently imposed additional restrictions in the bottom trawlers' activities. According to the EU Regulation 3699/93 the HP of the fleet trawlers must be reduced by 20 percent by the year 2000.

Given that HP has increased faster than the quantity of fish harvested, the catch per unit of this production input exhibits a clear downward trend during the period 1964-95. From more than 0.2 tons per HP in the 1960s and the early 1970s, it has fallen to less 0.16 tons per HP, on average, in recent years (Figure 9.3).

Figure 9. 3: Productivity of HP in the Bottom Trawler Fleet in Greece

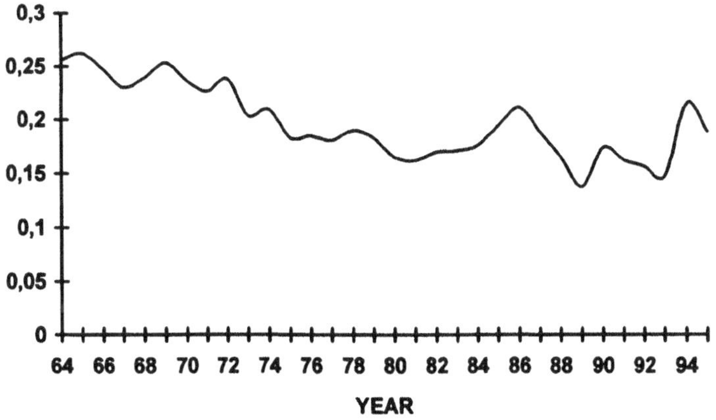

The Empirical Model and the Estimation Results

The empirical analysis of productivity growth requires knowledge of the fishing effort, E. In economics, effort is conventionally defined as an index of all inputs used for locating and harvesting fish (Hartwick and Olewiler, 1986). The construction of such an index requires knowledge of both the quantities and the prices of all inputs used by the fleet in question (Diewert, 1976). The prices and the quantities of most inputs used by trawlers are not available at the national level. The NSSG monitors only the number of vessels, the level of employment, the GRT and the HP of the fleet. Given the lack of the necessary data, the construction of an index for effort is not possible. Therefore, as in earlier empirical studies on the Greek fisheries (Katos, 1988; Malandrakis and Xepapadeas, 1992), this chapter uses a single input, namely the Horse Power, as a proxy for the fishing effort. We feel that the choice of the HP is absolutely justified on the grounds of the importance of this input for the fleet of bottom trawlers.

The system of equations (1)-(4) was estimated using the Kalman Filter Algorithm (Harvey, 1991; Hamilton, 1996) in the STAMP (Structural Time Series Analyzer, Modeller, and Predictor) program. This algorithm decomposes *y* into the permanent component, *g*, and the temporary component, *e*, that is, it performs 'signal extraction' on *y* (Enders, 1996). Table 9.1 presents the estimation results.

The stochastic shock, that is the term e in equation (2), appears to have the largest variance, followed by the stochastic shock to the level of the permanent component, that is the term *n* in equation (3). The

TABLE 9. 1: Estimation results

a) Variances of the Stochastic Shocks		
	Absolute Value	Ratio
S_e	0.0720	1.0000
S_n	0.0310	0.4330
S_z	0.0035	0.0480
b) Level of the Stochastic Trend, Elasticity of Harvesting with Respect to Effort		
	Level (1995)	r
	6.595	0.305
	(2.31)*	(0.193)*

PEV (1995):	0.08	
DW:	1.98	
Q(8,6):		$7.55 < (X^2_{0.05,8}) = 12.9$
BS:	$5.82 < (X^2_{0.05,2}) = 5.99$	
R_d^2 :	0.36	

- Standard errors in parentheses. Statistically significant at 5 percent level or less.

variance of the stochastic shock to the slope, that is the term *z* in equation (4), is the lowest. The variances of *n* and *z* are 43.3 percent and 4.8 percent of the variance of e, respectively.

Both the level of the stochastic trend, g, and the elasticity of harvesting with respect to effort are statistically significant at 5 percent level or better. The DW-statistic suggests that the estimated model is free of first-order autocorrelation. The portmanteau Ljung-Box Q-statistic suggests that there is no serial dependence in the first eight lags.[3] The R_d - statistic is a measure of goodness of fit computed on the differences.[4] It appears that the empirical model explains 36 percent more of the variability of the dependent variable than the basic alternative specification according to which the permanent component, g, follows a simple random walk. The Bownman-Shenton (BS) statistic (which measures the departure of the third and fourth moments of the stochastic shocks *e, n,* and *z*, from their expected values under normality) indicates that the assumption of normality cannot be rejected at the 5 percent level. The elasticity of harvesting with respect to effort (HP) is 0.305. This implies that the fleet experiences decreasing returns to scale with respect to this input.

Figure 9. 4: Rate of Change in the Stochastic Trend

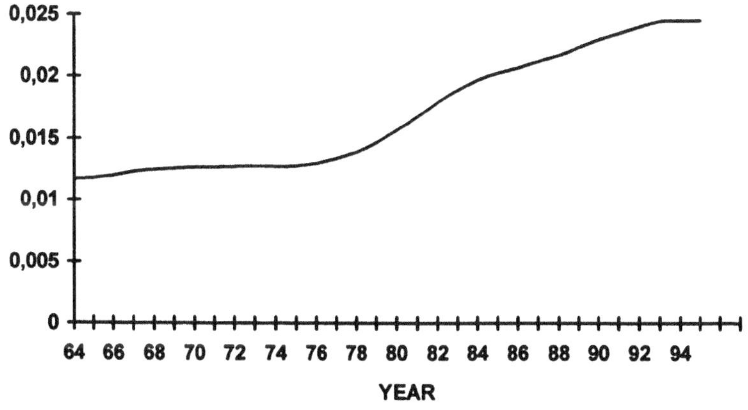

The slope (i.e., rate of change) of the stochastic trend for the period 1964-95 is presented in Figure 9.4. As explained earlier, the slope of g is the productivity effect, that is, the combined influence of technological innovations and permanent changes in fish stocks and their density on the productivity of effort. We observe that the productivity effect is positive everywhere. We also observe that the productivity effect remained almost constant from 1964-75 and afterwards it has increased rapidly to reach a rate 2.4 percent per annum in 1995. The productivity effect in the most recent years is more than two times higher than that prior to the mid-1970s.

Decomposition of Productivity Growth

The decomposition of fishing effort productivity of Greek bottom trawlers into a scale effect, a productivity effect, and a temporary effect is presented in Table 9.2. We have chosen to present average values for the periods 1964-83, 1984-95, as well as for the whole sample. The choice of the two sub-periods is not accidental. In October 1983, the EEC initiated its Structural Policy for the fisheries sector via regulations 2908/83 and 2909/83. In the context of this Structural Policy, grants were provided for the construction of new vessels and the modernization of existing ones. The analysis for the periods prior to and after the initiation of the Structural Policy is designed to determine the extent to which this policy had any influence on productivity and/or its components.

TABLE 9. 2: Productivity analysis

Period	Average Rate of Productivity Growth	Scale Effects	Productivity Effects	Temporary Effects
1964-83	- 0.0211	-0.0263	0.0136	-0.0084
1984-93	0.0082	-0.0260	0.0250	0.0092
1964-95	- 0.0097	-0.0262	0.0171	-0.0006

The fishing effort productivity of the Greek bottom trawler fleet fell at an annual average rate of 2.11 percent during the period 1964-83. This reduction must be attributed to the fact that a negative scale effect and, to a lesser extent a negative temporary effect outweighed a positive productivity effect. The negative scale effect was the direct consequence of the decreasing returns to scale with respect to effort. For the period 1964-93, the absolute value of the scale effect was almost two times the value of the productivity effect.

During 1984-95, fishing effort productivity of the Greek bottom trawlers increased at an average rate of 0.82 percent per annum. That increase must be, to a large extent attributed to the rise in the productivity effect compared to the period 1964-83. It is noteworthy that although the scale effect is almost identical in the two periods, the productivity effect in the latter period is 1.8 times higher than the productivity effect in the former one.

Over the whole period examined (i.e., 1964-95), the productivity of effort fell at an annual average rate of 0.97 percent per annum. The scale effect, with an average rate of change of -2.62 percent per annum, is the major determinant of productivity change, followed by the productivity effect (average rate of 1.71 percent per annum). The results suggest that, over 1964-95, 60 percent of the change in overall productivity can be explained by the scale effect, 39 percent can be explained by the productivity effect, and only 1 percent by the temporary effect.

Conclusions and Policy Implications

The objective of this chapter has been to investigate the fishing effort - harvest relationship and analyze the productivity of effort of the Greek bottom trawler fleet. For the purpose of the analysis a stochastic trend has been used to capture the impact of stochastic shocks due not only to technological innovations but also to fish stock density changes. The major findings stemming from the obtained empirical results may be outlined as follows:

The fleet of bottom trawlers in Greece experiences decreasing returns to scale with respect to effort (which in the present chapter has been identified with HP). The decreasing returns to scale are, to a large extent, responsible for the fall in the productivity of the fleet's effort during the period 1964-95.

An elasticity of harvesting in the range of 0.3 implies that substantial reductions in HP are required to achieve a sizeable reduction in harvesting. As mentioned in part 3, the EU Regulation 3699/93 stipulates a 20 percent reduction in the HP of the fleet by the year 2000. Even if Greece achieves this objective -- a rather bleak prospect given the objections of the vessel owners and the concerns of the Greek authorities about the socio-economic impact of such a measure in the coastal areas of the country (Operational Program of Greece for Fisheries, Ministry of Agriculture, 1994) -- the expected reduction in harvesting will be in the range of 6 percent only.

However, the productivity effect has, to a certain extend, offset the negative scale effect, and a strong overall productivity effect is observed after the early 1980s. As technological innovations are typically embodied in new capital goods, this finding provides an indication that the financial incentives offered in the context of the EU Regulations 2908/83 and 2909/83 (pertaining to the construction and modernization of fishing vessels) worked towards an increase in the fleet's overall productivity.

The average productivity growth over the period 1964-95 has been 1.71 percent per annum. However, the policy implications one may draw from this result depend on the assumptions one is willing to make about the nature of the factors that have brought about this productivity effect. As mentioned in part 1, the stochastic trend may capture not only technological innovations but also changes in the level and the density of the fish stocks. Conceivably, three alternatives can be considered regarding the factors responsible for the measured productivity of bottom trawlers. Under the first alternative, the average shocks in the level and density of the fish stocks are zero. Then, the productivity effect is in the form of a purely technical change effect. Under the second alternative, the average shocks in the level and the density of the stocks are positive. Then, the productivity effect overestimates the technical change effect. Under the third alternative, the average shocks in the level and the density of stocks are negative. Then, the productivity effect underestimates the technical change effect.

Given the large number of fish species in Greek waters there is no reliable information about the density and the level of fish stocks (Operational Program of Greece for Fisheries, Ministry of Agriculture 1994; Karagiannakos, 1994; European Union Report, 1994). The general belief, however, is that some species are over-exploited, which means that their stocks and density are likely to be decreasing, while some others are moderately exploited. Provided that this general belief reflects reality about fish stocks in Greek waters, the productivity effect found in this chapter is likely to underestimate the rate of technological change in the bottom trawler fleet.

Overall, two major conclusions may be drawn from the preceding discussion. First, if indeed fish stocks are being significantly depleted in Greek waters, then the underestimated rate of technical change and the low elasticity of harvesting with respect to HP clearly point toward a rather important policy implication: they constitute evidence that the EU and the national fishing authorities should focus on policy measures discouraging the use electronic equipment and other advanced fishing gear, i.e., capital goods embodying the bulk of technical innovations in bottom trawler fishing in order to achieve the goal of fish stock preservation.

Moreover, the inclusion of an environmental dimension such as permanent changes in the fish stocks, although indirect in our approach, points out the importance that such factors may potentially have on key measures of a sector's performance. It also stresses the often pressing lack of specific information on the mechanisms of such biological processes, which makes it infeasible to account for it in empirical economic research. A priori information from other disciplines that may lead to explicit modeling of such factors may significantly contribute to an enrichment of conventional economic measurements and provide an excellent opportunity for interdisciplinary research.

Notes

[1] From assumptions (a) and (b) it follows that the stochastic harvest-effort relationship employed in this chapter is similar to the deterministic one of the form $Q = E^{r}\exp(at)$ which is widely used in empirical studies of the fisheries sector (e.g., Bjorndal and Conrad 1987; Conrad and Clark, 1991; Malandrakis and Xepapadeas, 1992).

[2] The standard decomposition of productivity growth involves scale and technical change effects (e.g., Chambers, 1989; Glass and McGillop, 1992). In our chapter, however, the permanent component captures both technological innovations as well as permanent changes in the density and the level of fish stocks. To distinguish changes in g from pure technical changes effects, we use the term 'productivity effects' suggested by Harvey (1991).

[3] The Q-statistic is calculated as $Q(p,d) = T(T+2)\sum_{j=1}^{p}\frac{r_j^2}{T-j}$, where p is the number of lags, d=p-s+1, s being the number of stochastic shocks (in this chapter there are three, e, n, z), r is correlation coefficients, and T the sample size. The Q-statistic is distributed as a X^2 with d degrees of freedom.

[4] The coefficient of determination, R^2, is not a suitable measure of goodness of fit when the variables in model exhibit trends (Maddala, 1992). Since this is the case in the present chapter, we use the alternative measure

$R_d^2 = 1 - \frac{(T-d)PEV}{\sum_{t=2}^{T}(\Delta Q - \overline{\Delta Q})}$, where PEV is the one-step ahead prediction error variance at the sample end.

References

Anderson L. 1985. *The Economics of Fisheries Management*, Baltimore, MD: The Johns Hopkins University Press.

Bernard, A. and C. Jones, 1996. « Productivity Across Industries and Countries: Time Series Theory and Evidence,» *Review of Economic and Statistics,* LXXVIII: 135-46.

Bjorndal, T. and J. Conrad, 1987. «The Dynamics of an Open Sea Fishery,» *Canadian Journal of Economics,* 1: 74-85.

Chambers, R. 1989. *Applied Production Analysis: A Dual Approach,* New York: Cambridge University Press.

Clark, C. 1985. *Bioeconomic Modelling and Fisheries Management,* New York: John Wiley & Sons.

Clark, S. and C. Youngblood, 1993. «Estimating Duality Models with Biased Technical Change,» *American Journal of Agricultural Economics,* 75: 353-60.

Conrad, J. and C. Clark, 1991. *Natural Resource Economics: Notes and Problems,* Cambridge: Cambridge University Press.

Diewert, E. 1976. «Exact and Superlative Index Numbers,» *Journal of Econometrics,* 4: 115-145.

Enders, W. 1995. *Applied Econometric Time Series,* New York: John Wiley & Sons.

European Union Report, 29/8/1994.

Glass, J. and D. McGillop, 1990. «Production Interrelationships and Productivity Measurement in Irish Agriculture,» *European Review of Agricultural Economics,* 17: 271-87.

Hamilton, J. 1996. *Time Series Analysis,* Princeton: Princeton University Press.

Harvey, A. 1989. *Forecasting, Structural Time Series Models and the Kalman Filter,* Cambridge: Cambridge University Press.

Hartwick, J. and N. Olewiler, 1986. *The Economics of Natural Resource Use,* New York: Harper and Row.

Karagiannakos, A. 1994. *Fisheries Management in the EU,* Aldershot: Avebury.

Katos, A. 1988. «Productivity and Resource Stocks in Greek Fisheries,» *Review of Agricultural Studies,* 1: 45-54 (in Greek).

Maddala, S. 1992. *Introduction to Econometrics,* Englewood Cliffs, New Jersey: Prentice Hall.

Malandrakis, G. and A. Xepapadeas, 1992. «Production Analysis in the Open Sea fishery of Greece,» paper presented at the 2nd Panhellenic Meeting of Agricultural Economics, Thessaloniki, 11-12 Dec.

Operational Program of Greece for Fisheries, Ministry of Agriculture, Oct. 1994 (in Greek).

Pindyck, R. 1984. «Uncertainty in the Theory of Renewable Resource Markets,» *Review of Economic Studies,* 51: 289-303.

Squires, D. 1987. «Fishing Effort: Its Testing, Specification, and Internal Structure in Fisheries Economics and Management,» *Journal of Environmental Economics and Management,* 14: 268-82.

Chapter 10

Exploring the Sustainability of Aquaculture

Giannis Karagiannis, Stylianos Katranidis, Elisavet Nitsi, Evangelos Tzouvelekas, Alexandra Vakrou

Introduction

In the case of fisheries, international trade is blamed for contributing to overfishing, and although for many fisheries the threat of biological extinction does not arise, commercial collapse and losses in genetic and biological diversity may interfere with the long run sustainability of the resource (Sen, 1994). Unlike fisheries, aquaculture farms constitute a major factor towards avoiding resource depletion and resource scarcities, by producing and marketing perfect substitutes. However, production is associated with problems of environmental quality which may pose a threat to marine ecosystems and local quality of life.

During the last decade, aquaculture has undergone a rapid development in Greece and has become one of the most significant sectors of primary production, with an important contribution both to primary sector GDP and exports. The sector's position and prospects also appear to have great importance at the international level and especially in Europe, where aquaculture exhibits high development rates, and Greece holds to date the first place in production volume and exports.

In the light of the need for liberalized policies in the primary production sector, there are two important aspects related to the sustainability of aquaculture which require examination. The first aspect refers to the development prospects of the sector, since the international environment appears to be highly competitive and the sector has a definite export based profile. Thus, the examination of productivity and economic efficiency, two of the most important elements for the formation of competitiveness, constitute the first set of issues to be analyzed in the context of economic sustainability. The second aspect is related to both the environmental and social sustainability of aquaculture[1]. The sector constitutes a source of income and employment, but it also generates negative environmental impacts which have become a front line issue. Thus, the social valuation of any possible environmental impacts will provide some evidence on the social acceptability of this activity.

The analysis of these issues is approached using different procedures. More specifically, for the estimation of the sector's competitiveness and productive efficiency, a sample of 10 enterprises spread throughout Greece has been used, ; while for the estimation of the environmental impacts of aquaculture, the installations in the island of Kefalonia, in the Ionian Sea, have been researched as a case study.

Competitiveness and Efficiency

Productive efficiency is composed of technical and allocative efficiency. Technical efficiency refers to the production at given output levelS with the smallest possible input. For a given input, a lower output level than is technically feasible, is evidence of inefficiency. By improving the current know-how, output increased without any cost increases. Allocative efficiency refers to the achievement of least cost in producing a given output level. This means an input combination that ensures cost minimization. Productive efficiency is measured by the product of technical and allocative efficiency. Productive inefficiency implies extra cost and reduces a firms' competitiveness, as it has to charge a higher price for its products in order to cover average cost.

Technical efficiency measurement is based on stochastic frontier function methodology, and, in particular, the generalized framework developed by Battese and Coelli (1992). Application of this framework to cross-section data requires particular restrictions in the specification of the density function corresponding to the stochastic term associated with technical inefficiency. This stochastic term follows a truncated distribution with positive mean and unknown, but finite variance. Measures of technical efficiency are obtained through the estimates of the mean and the variance of truncated distribution. For the purposes of the present study, these estimates are obtained by using the method of maximum likelihood, as it is provided by the FRONTIER computer program developed by Coelli (1992).

The choice of a functional form to approximate the underlying production function is important for the measurement of allocative efficiency. According to Bravo-Ureta and Rieger (1991), self-dual functions have the advantage of providing measures of both technical and allocative efficiency by single-equation estimation of the production frontier function only. For this reason, a Cobb-Douglas production frontier function has been used in this study, with a Divisia aggregator as an output index and three factor inputs, i.e., fry, feeding and labour. The estimated production frontier function is given as:

$$\underset{}{lnQ} = \underset{(0,22)}{0,84} + \underset{(0,04)}{0,03\, lnX_1} + \underset{(0,07)}{0,79\, lnX_2} + \underset{(0,02)}{0,04\, lnX_3} \qquad (1)$$

where Q refers to output, X_1 to fry, X_2 to feeding and X_3 to labor. Numbers in parentheses are standard errors of the estimated parameters. Using standard duality results, the corresponding cost function is given as:

$$lnC = 0,88 + 1,16\, lnQ + 0,03\, lnW_1 + 0,92\, lnW_2 + 0,05\, lnW_3 \qquad (2)$$

where W refers to input prices.

Returns to scale can be measured directly through the estimated parameters of either the frontier production or the cost function. In the former case, they are measured by the sum of production elasticities (e.g. 0.03+0.79+0.04), and in the latter by the inverse of the *lnQ* coefficient. Returns to scale were found to be decreasing in the Greek aquacultural sector being 0.86 on average. This means that 1% increase of output results in a (1/0.86=) 1.16% increase of total cost.

Measures of productive efficiency are reported on Table 10.1. Technical efficiency has been measured by following Battese and Coelli (1992) methodology, while that of allocative efficiency has been based on Bravo-Ureta and Rieger (1991) method. Table 10.1 presents the frequency distribution of technical, allocative and productive efficiency. Mean technical efficiency WAS found to be around 67%, which means that a 33% increase of production is feasible with present technology and unchanged input quantities, as long as producers know-how is improved. This may be due to the infant stage of the sector, which started its operation in the middle '80s. Nevertheless, most firms in the sample (70%) have achieved technical efficiency greater than 60%.

Mean allocative efficiency was found to be around 80%, which is higher than technical efficiency. This means that firms have achieved a relatively better allocation of existing resources, but still, a 20% decrease in cost is feasible by a further re-allocation of inputs for any given level of output. The frequency distribution results show, however, that around 75% of the firms in the sample achieved more than

70% of allocative efficiency. Based on the above estimates, mean productive efficiency is found to be around 53%, mainly due to the low degree of technical efficiency.

TABLE 10. 1: Degree of technical, allocative, and economic efficiency for aquaculture farms in Greece

Efficiency Level (%)	Technical Efficiency	Allocative Efficiency	Economic Efficiency
<20	0	0	0
20-30	0	0	2
30-40	4	1	7
40-50	4	1	8
50-60	4	3	9
60-70	7	5	8
70-80	11	7	3
80-90	7	11	2
90-100	3	12	1
Number	40	40	40
Mean	67,22	80,15	53,48
Minimum	31,92	36,33	28,23
Maximum	96,98	98,58	90,73

It is apparent that there is considerable scope for improvement in the efficiency of production for Greek aquacultural firms, which would thus achieve significant cost savings. According to our results, more savings are expected by the improvement of technical efficiency. This can be achieved through learning-by-doing economies, which are expected in the near future as the sector is still at an infant stage. Consequently, its competitiveness will increase, resulting in lower consumer prices and expanding exports.

Environmental Impacts

Several methods have been developed for estimating values for different environmental impacts. These methods differ significantly from each other, in their theoretical background, as well as in the estimation techniques they employ and the difficulty in their application (Hanley and Spash, 1995).

For valuing the environmental impacts of aquaculture in the island of Kefalonia, we use the contingent valuation method (CVM), mainly because this method can be adapted for the valuation of most environmental goods or services, and does not require the availability of large quantities of data (Mitchell and Carson, 1989). In parallel, the questionnaire used was designed to collect the necessary information for the application of travel cost method (TCM), if this would be permitted by the findings of the survey (i.e. if notable changes in the use patterns of the area for swimming were recorded).

Impacts of aquaculture

The study is focused on the identification and analysis of the impacts of aquaculture development in the island of *Kefalonia*. We chose to employ a questionnaire based study of the local population (both residents and tourists have been interviewed), in order to gather information on changes in level of use of the area where the aquaculture units are. This would reveal people's willingness to pay for closing down the units, or their willingness to accept compensation in order to accept the operation of the plantations; and finally to identify people's attitudes towards the aquacultural development of the area.[2] Besides the fact that on site research of the physical and technical parameters of the aqua-culture and the marine environment has revealed that there are no important impacts from the establishment of the aquaculture units, the attitude of the people in the island has two very distinct dimensions. Some people are against aquaculture, while some others view it very positively, thinking that it is beneficial for the island and they would be willing to support its further development. The study analyzed the spatial differentiation of the answers to questions aimed at identifying people's perception of various impacts from the development of aquaculture, as well as differentiation according to the nationality of the respondents and their place of residence.

The questionnaire included a set of questions aiming to identify the use of the aquaculture area for fishing and the fishermen's previous and current level of use of it. Research has revealed that there is no decline in the number of marine species in the area, and information from the scientists of the units, suggest that the fish population in the area might have increased, as it is enhanced by fish from the aquaculture pods. The data in Table 10.2 indicates that 30% of the respondents fish in the area. There is also a considerable difference in the respondents' reactions regarding their judgement of the area as a fishing place: 55% of them indicated that the area is a good fishing place, 30% that it is average, while 15% think that it is a bad place. There were no respondents who used to fish in the area in the past but do not at present due to a decline in its capacity. It is important that the majority of respondents (90%) stated that the condition of the area before and after the creation and the operation of the aquaculture units was the same, while only 3% indicated a decline and 7% an increase in its fishing quality.

TABLE 10. 2: Use and valuation of the area as a fishing place

		Valuation of the area as a fishing place		
Use of the area for fishing	Number (%)	Good	Average	Bad
Yes	80 (30%)	44 (55%)	24 (30%)	12 (15%)
No	185 (70%)	-	-	-
Total	265 (100%)			

The area is also used for swimming. In this case 40% of the respondents indicated that they swim in the area and 30% of these did not indicate any other alternative area. In Table 10.3 the relation between the use of the area for swimming and the distance between the residence of the respondent and the area is presented. Respondents staying in the village of *Farsa*, where the aquaculture units are located, use the area exclusively, while residents of *Argostoli* and *Lixouri*, the two biggest towns of the island, as well as of *Petrikata*, had the lowest percentages for using the area for swimming.

This evidence suggests that using the area for swimming depends on the distance of the respondent's residence from the area, the only exception being the residents of Argostoli. This can be explained by the fact that there are a lot of other more beautiful beaches near this town. Respondents who stated that they do not use the area for swimming, explained their attitude was a consequence of the distance from their residence (32%), the bad quality of the area (31%) and the lack of good access to it (8%). It can be concluded that the sea area of the aquaculture units has not been of a high swimming potential and thus the development of the units did not interfere with the swimming patterns of the residents in the island.

TABLE 10. 3: Respondent's swimming behavior by place of residence

Place of Residence	Distance (km)	Not Swimming %	Swimming %
Argostoli	8	93	7
Lixouri	21	89	11
Riza	7	14	86
Kardakata	8	29	71
Petrikata	9	86	14
Farsa	1	0	100
Kourouklata	4	50	50
Livadi	15	25	75
St. Dimitrios	17	39	61

χ^2 = 122.4959 (Significant at .05% level)

The questionnaire employed several statements which identified certain impacts of aquaculture: positive, like employment opportunities for the islanders and supply of fish for the dietary requirements; negative, like sea pollution, decline of landscape aesthetics, disturbance from the noise and the traffic around the units; and, one neutral that stated that aquaculture causes no impact at all. The respondents were asked to mark these statements on a scale of 1 to 5, with 1 being the lowest impact and 5 the greatest. Tables 10.4 and 10.5 present the results from the analysis of variance between people's responses by their place of residence as well as the responses of local inhabitants and Greek and foreign tourists.

The specific results in Table 10.4 indicate that there are significant differences at a 5% level regarding the importance of aquaculture for the creation of new jobs in the island, the decline in the landscape aesthetics and the pollution of sea, while there were significant differences at the 10% level for the statements regarding the importance of the farms for the production of fish for the population of the island, as well as for the statement of no disturbance.

TABLE 10. 4 : ANOVA for respondents' answers on environmental statements by place of residence

	New jobs	Supply of fish	landscape aesthetics	Sea Pollution	No Disturbance
Farsa	4.0000	2.3333	1.4074	2.3704	1.5380
Wider area	4.9459	2.9167	1.0263	2.7027	1.0263
Area across	4.5619	2.3962	1.0357	1.0179	1.0000
Argostoli	4.6102	2.7544	1.0179	1.2321	1.0000
Lixouri	4.8154	2.5574	1.0625	1.6984	1.0000
All respondents	4.6393	2.6017	1.0788	1.6611	1.0203
F - statistics	10.5272	1.9651	4.4754	31.4701	1.9892
Significant level	0.05%	0.10%	0.05%	0.05%	0.10%

More specifically, the statement regarding the importance of the farms in creation of jobs was found to be more important for the people living in villages around the farms. This finding was somewhat unexpected, due to the fact that residents of these areas do not work in the farms and thus they do not directly gain from them. High rates have also be given by residents of the two main towns of the island, Argostoli and Lixouri. The residents of the villages lying on the other side of the units, where the aquaculture employees, gave high rates too, but not as high as they should have given. The residents of Farsa, where the farms are located, gave the lowest rates.

To explore further the importance of the farms in creation of jobs, regional multipliers have been estimated for Kefallonia island. According to our estimates, the regional multiplier has been estimated to be 2.11[3]. This means that the creation of one income unit by one of the export sectors of the region, contributes to the creation of 1.11 additional income units in the economy of the region. Especially for the aquaculture sector, which has inputs amounting to 35% and exports to 95% of the production value, an increase of the production value by 1 unit leads to an increase of the regional income by 0.68 units and to an increase in the aquaculture income by 0.65. It should be mentioned that from the total primary sector production of the island, only the viticulture, olive tree cultivation, honey production and aquaculture contribute to the region's exports. All other activities, despite the fact that they represent 61% of the total agriculture and livestock production of the region, produce for the local economy and self-consumption. Regarding the manufacturing sector, only olive mills and wineries contribute significantly in the regional exports, while the rest are clearly orientated towards the local markets. Of the remaining sectors, only public administration and security, and the services, including hotels and restaurants, have been included in this analysis for the export sectors of the region.

On the other hand, sectoral income transfers to the island may also be taken into account when estimating regional multipliers[4]. For 1994, the sum of these transfers was 26.8 billion drachmas, an amount equal to 68% of the Gross Product of the prefecture for the same year. If we include this income in the export base income of the prefecture and more specifically during the estimation of the multiplier, then its value is 1.45, lower than the one estimated without taking into account these income transfers. That means that in case we want to include these transfers in our calculations, an increase by 1 unit of the aquaculture income, leads to an increase of the total income of the prefecture by 1.45 units. This multiplier, after the appropriate transformations and bearing in mind the economic data of the sector (i.e., inputs and exports of 35% and 95% of the production value respectively), indicates that an increase in the production value by 1 unit leads to an increase of the prefecture's income, irrespective of the aquaculture sector, by 0.28 units.

Regarding sea pollution, the highest values have unexpectedly been given by the residents of the surrounding villages, which are mountainous and thus not affected by any pollution. Residents of Farsa gave the second highest response, followed by residents of Lixouri and Argostoli, while the residents of the villages lying on the other side of the farms, who also happen to work in the farms, gave the lowest responses.

The residents of Farsa are the ones who have the most serious problems regarding the decline of the landscape quality and aesthetics of the area due to the farms. However, the same people indicated that they are not disturbed by the farms. Residents in the two towns of Argostoli and Lixouri valued the farms very highly for their contribution to the supply of fish to meet the island's needs.

The creation of new jobs was far more important for Greeks, irrespective of them being or not being permanent residents of the island.

The same reaction also holds true regarding sea pollution, an impact that foreign visitors in the island did not perceive as that important. The responses to statements regarding the impacts on the aesthetics and the importance for the production of fish for the island, as well as the statement concerning the disturbance from the farms did not vary significantly for these groups of respondents (Table10.5).

TABLE 10.5 : ANOVA for respondents' answers on environmental statements by relation with the island

	New jobs	Supply of fish	landscape aesthetics	Sea Pollution	No Disturbance
Residents of Kefalonia	4.6950	2.7090	1.0812	1.6990	1.0250
Greek tourists	4.7931	2.5714	1.2069	1.8571	1.0000
Foreign tourists	4.1290	2.4815	1.0000	1.0357	1.0000
All respondents	4.6385	2.6680	1.0853	1.6429	1.0190
F - statistics	11.0925	0.5852	1.5286	6.1460	0.2307
Significant level	0.05%	N. S.	N. S.	0.05%	N. S.

Evaluation of environmental impacts

When applying the CVM in the case of Kefalonia, both willingness to pay and willingness to accept compensation scenarios were used, following the 'take-it-or leave-it' approach, which asks questions with prefixed prices. However, several problems arise with the application of this method resulting in failure to estimate a value for the environmental impacts of the units. The reasons for these problems are the following:

- The nature of the environmental good in question is such that it does not permit an easy and accurate evaluation. The area where aquaculture was established is privately owned, and furthermore, this area was not used in the past for swimming or fishing by locals or tourists. Thus the environmental good in question (i.e. the area of the units) did not have any direct use in the past or at present. Problems of this type have also been encountered with previous CVM studies in similar cases (Hausman, 1995).

- It was very difficult to present the hypothetical market to the respondents and to communicate it effectively. Thus a crucial problem ariseS, that of understanding properly the good under investigation.
- The respondents to the CVM questions found difficulty in understanding in which way they would receive compensation from the owners of the aquaculture units, since the area of the units is privately owned and these people have made huge investments for these installations that benefit the island by providing jobs to more than 100 people. When asked for their willingness to pay for compensating the entrepreneurs for the closing down or relocation of the units, this was also negative and it appeared to have two very distinct motivations. Some people asked for the development of more units, took a strongly positive position towards this kind of development, accepting it as beneficial for their island. Others had objections and stated that the units do cause environmental damage, but they were not willing to pay anything for the relocation of the units, claiming that the state is responsible for the quality of the environment and should, therefore, compensate the owners of the aquaculture. This same reaction is quite common in other CVM studies where respondents gave protest bids with the same explanation (Papageorgiou and Vakrou, 1997).

However, although the application of the CVM in the Kefalonia survey was not fruitful for all the reasons mentioned above, the survey did gather information on the people's attitudes towards the aquaculture installations and this form of development.

Interpretation of respondents' perceptions of aquaculture

A further research objective was the investigation of the possibility to forecast the attitude of the respondents according to their particular socio-economic characteristics and the level of disturbance from the units. To achieve that a logit model was applied to the data, in which the dependent variable is a dummy, taking the value 0 if the attitude is positive and 1 if the attitude is negative.

The probability of a negative attitude is estimated by:

$$Probability_{(negative\ attitude)} = 1 / (1 + e^{-z}) \qquad (3)$$

In this equation z is estimated by :

$$z = \beta_0 + \beta_1 {*} X_1 + \beta_2 {*} X_2 + \ldots + \beta_k {*} X_k \qquad (4)$$

where: β_0, β_1, β_2, ..., β_k are the coefficients of the logit equation and X_1, X_2, ..., X_k are the variables taking part in the logit equation.

Thus the z can be calculated for each respondent, by substituting the values for each one in each respective variable in the model. Then the probability of negative attitude is estimated by entering z into equation (4). If the price estimated for the probability is $\geq$ than 0.5, then the particular respondent will be stated to have a negative attitude.

The application of this model leads to the creation of an equation which:

- identifies the important variables which determine the attitude of the respondent
- permits the estimation of the attitude of any other respondent by using the model estimated and entering his own socio-economic information into it.

For the estimation of the logit model, 15 variables were used, 10 of which represented socio-economic variables of the respondents and 5 their personal score for the statements regarding the environmental impacts of aquaculture in the area. To enter the model, a variable needs to satisfy the Likelihood Ratio. The results of the logistic regression are presented in Tables 10.6 and 10.7.

Table 10.6: Results of the logistic regression model

		χ^2	Degrees of Freedom	Significant
-2 Log Likelihood		145.627	178	0.9638
χ^2 of the model		109.256	5	.0000
Goodness of Fit		10561.246	178	.0000
Variable	B	S.E.	Wald test	Sign.
Importance of fish production	-0.7988	0.2180	13.4241	0.0002
Impact on landscape and aesthetics	-4.1447	8.2642	0.2515	0.6160
Sea Pollution	-2.5728	0.4024	40.8855	0.0000
Dummy (local-tourist)	-0.9776	0.3183	9.4356	0.0021
Dummy (resident of the villages where employees of the units stay	-3.1186	0.6553	22.6470	0.0000
Constant	10.9547	8.3575	1.7181	0.1899

Table 10.6 presents the logistic regression and the information regarding its goodness of fit and its chi-square. The variables entered in

the final equation were the score for the importance of the aquaculture units for the production of fish, the pollution of the sea, a dummy identifying if the respondent is a permanent resident of the island or a tourist, a dummy for respondents staying in villages whose residents are employed in the aquaculture units, and the impact on the aesthetic of the area.

The equation indicates that the most important variables which influence the attitude of any respondent towards the aquaculture units and development are the ones which have the greatest values in the equation, i.e. the dummy indicating if the respondent lives in a village the residents of which are employed in the units, and the score for the pollution of the sea. Due to the values that the particular variables take, an increase in their value corresponds in an increase of the probability that the respondent will have a negative attitude towards aquaculture. Although the variable for the respondent's score for the impact on aesthetics was included in and improved the equation, at the end it does not appear to be no statistically significant and thus cannot be accepted as a factor for the identification of respondent's attitude towards aquaculture.

Table 10. 7: Prediction of attitudes towards the development of aquaculture

	Predicted values		Percentage of responses predicted correctly
Observed values	Positive Attitude (0)	Negative Attitude (1)	
(0)	80	15	84.21%
(1)	11	78	87.64%
Overall prediction			**85.87%**

Table 10.7 presents the projection of attitude estimated by the equation (1=negative, 0=positive). The equation has a very good fit to the data, since it is able to predict accurately 86% of the sample answers.

Conclusions

The efficiency analysis conducted for 40 enterprises of the aquaculture sector show that there are considerable chances of improving the efficiency of the bass and bream aquaculture, thus the reduction of the operation costs are expected to be high. More benefits are expected as a result of the improvement of the production technology, as the degree of technical efficiency was found to be low. That means that there are still very good chances for the improvement of the enterprises' overall competitiveness.

The information collected with an on-site survey in the island of Kefalonia revealed that the establishment of aquaculture units in the island have not affected the use of the area for fishing by the local people. In the same way, its use as a swimming area has not been affected either, since in the past the area was not popular for this type of use and if somebody wished to use it, they still could.

The importance of the farms in creation of jobs was found to be more important for the people living in villages around the farms and high rates have also been given by residents of the two main towns of the island, Argostoli and Lixouri, while residents of Farsa, where the farms are, gave the lowest rates. Regarding the pollution of the sea resources, the highest values have been given again by the residents of the surrounding villages, while residents of Farsa gave the second highest responses, followed by residents of Lixouri and Argostoli, and the lowest responses have been given by the residents of the villages lying on the other side of the farms and they work in the farms.

The residents of Farsa are the ones who have the most serious problems regarding the decline of landscape quality and aesthetics of the area due to the farms. However the same people indicated that they are not disturbed by the farms. Residents in the two towns of Argostoli and Lixouri valued the farms very highly for their contribution to the supply of fish for the island's needs.

The impact of job creation, together with the pollution of water resources were far more important for Greeks, irrespective of them being or not being permanent residents of the island. Answers to the statements regarding the impact on the aesthetics and the importance for the production of fish for the island, as well as to the one concerning the disturbance from the farms did not vary significantly between respondents.

People are more likely to develop a positive attitude towards aquaculture farms if they are permanent residents of the island or visitors, staying in a village whose residents are employed in the farms, if they believe that the farms contribute to the creation of new jobs and if they do not believe that the farms pollute the water resources.

All the previous findings of this multidisciplinary approach indicate that aquaculture is a very promising form of development, which is particularly suited to the Greek island and seaside areas. The analysis of the efficiency of the units shows that aquaculture is economically viable, but most importantly that there are opportunities for further development. Also aquaculture is very important for the island areas because it provides a lot of job opportunities and in this way can minimize whatever negative attitudes exist in the local societies. If, in addition, the choice of the area where the farms are going to be established will be made with care, and attention is paid so that any environmental impacts are minimal or possibly nil, then it can be concluded that this form of development is one that can meet sustainability criteria.

Notes

[1] In many cases socio-cultural sustainability denotes the reduction of social conflict (UNEP *et al.*, 1991; Kousis, 1993).

[2] The survey took place in the wider area of the aquaculture units during July and August 1997. The design of the questionnaire followed the theoretical framework and the usual guidelines for surveys of the kind and gathered information on the socio-economic profile of the respondents, their attitude towards aquaculture, their opinion about the specific units and their patterns of using the area (Oppenheim, 1992). The area covered by the survey was restricted to the west part of the island, since the existing aquaculture farms affect only the people living at or visiting this part of the island.

[3] Our analysis uses the export base approach (Pleeter, 1980). The reason for this choice was that the approach which is based on the intersectoral analysis and the input - output technique is meaningful when in the economy under study important intersectoral exchanges are taking place, as is the case in the manufacturing sector. But this situation cannot be observed in the island of Kefalonia, as well as in areas with an agriculture based economy, where the leading economic sectors are primary production and services. On the other hand, the aquaculture sector is not characterized by strong links with other sectors, especially at an interregional level. At this point, it has to be noted that the aquaculture sector has input expenditures of around 35% of the production value, while in the case of 'Aquacultures of Kefalonia' the relevant percentage of interregional inputs does not exceed 8%. Another reason for choosing the export base approach lies in the availability of the necessary regional data. The export base model is in general less data demanding and the only data required is regional exports by sector. On the other hand for the application of the intrasectoral analysis, apart of the aforementioned information, a full estimation of the trade balance by sector as well as the regional sectoral production functions, would have been required.

[4] Special characteristics of the economy of the island, like the importance of the agricultural employment and the respective income transfers and subsidies, the increased employment of the islanders in shipping, the relatively big immigration, as well as the crucial role of social income transfers and especially the pensions, lead us to the estimation of the respective income transfers.

References

Battese, G. E. and Coelli T. J. 1992. «Frontier Production Functions, Technical Efficiency and Panel Data with Application to Paddy Farmers in India,» *Journal of Productivity Analysis*, 3: 153-169.

Bravo-Ureta, B. E. and Rieger, L. 1991. «Dairy Farm Efficiency Measurement Using Stochastic Frontiers and Neoclassical Duality,» *American Journal of Agricultural Economics*, 73: 421-428.

Coelli, T. J. 1992. «Computer Program for Frontier Production Function Estimation, FRONTIER, Version 2.0,» *Economics Letters*, 39: 29-32.

Hanley, N. and C. L. Spash, 1995. *Cost-Benefit Analysis and the Environment*, Cheltenham: Edward Elgar.

Hausman, J. A. (ed.) 1993). *Contingent Valuation: A Critical Assessment*, Contribution to Economics Series, 220, Elsevier Science Publishers, The Netherlands.

Kousis, M. 1993. «Collective Resistance and Sustainable Development in Rural Greece: The Case of Geothermal Energy on the Island of Milos,» *Sociologia Ruralis*, 33(1): 3-24.

Mitchell, R. C. and R. T. Carson, 1989. *Using Surveys to Value Public Goods: The Contingent Valuation Method*, Washington DC: Resources for the Future Inc.

Oppenheim, A. N. 1992. *Questionnaire Design, Interviewing and Attitude Measurement*, London: Printer Publishers Ltd.

Papageorgiou K. and A. Vakrou, 1997. «Estimating the benefits from recreation in two Greek national parks: the application of the contingent valuation method,» paper presented in the European Environmental Research Network Conference, Athens 15 February 1997.

Pleeter, S. 1980. *Economic Impact Analysis, Methodology and Applications*, Studies in Applied Regional Science, No. 19, Boston.

Sen, S. 1994. «The Environmental Effects of Trade in the Fisheries Sector,» in *The Environmental Effects of Trade*, OECD, Paris.

UNEP, IUCN, and WWF, 1991, *Caring for the Earth*.

Chapter 11

Trade, Sustainability, and the Forest Sector

Nikolaos Stamou

Introduction

After its initial coining, almost three centuries ago, by the forester Carlowitz (1713), sustainability has acquired an evolutionary meaning, related to the needs of human society which the forests are supposed to meet, and to the way it had been applied later (DPIE, 1996). For many years, the application of sustainability has been focused on timber cuttings, since the main interest of society was exclusively orientated toward wood products. However, in the last decades, the meaning of forest sustainability has been transformed to connote a balance in the forest's multiple functions. These functions are related to the three basic uses of forests, that is the consumptive uses (e.g., harvesting and hunting forest resources, recreational uses), the non-consumptive uses (e.g., ecotourism, genetic resources and existence values) and the ecological functions (watershed protection, carbon storage, and microclimatic role) (Barbier, 1994).

Any interference in one or more of the aforementioned specific functions of a forest constitutes a disturbance to the existing balance. In the case of interference which affects negatively the protective-anti-erosion role of the forest, the soil's productivity is degraded, and this automatically reacts negatively also with all the rest of the functions of the forest, including timber production (Dafis, 1986). In other words, the display of the multifunctional role of forests constitutes a complex form of a joint production. So, the remark made by Carlowitz becomes opportune: «where damages occur on the soil due to omissions or exaggerations this is where poverty and deficiency occur» (Stamou, 1997).

Considering the multifunctional role of forests, a sustainable forest management cannot be focused on achieving targets which are connected with some partial functions of an area's forests (e.g. some consumptive and/or non-consumptive), without ensuring at the same time that the rest of the functions will continue to be fulfilled by the forest and be expressed quantitatively at the above minimum level. Along these lines, timber production from sustainable management and the marketing of forest products, under a freer international trade regime, may favor those countries in which the ecological functions of forests and ecosystem stability are affected less by this specific management.

Free trade cannot be judged *a priori* as beneficial or detrimental to forest sustainability. In Sub-Saharan Africa the shift in forestry towards exports may have been environmentally beneficial (Repetto, 1989). For export oriented tree crops, in West Africa, soil erosion tends to be three times less than crops aimed at domestic markets. In Sudan, however, trade and other reforms led to high rates of deforestation by encouraging woodland clearing for crop cultivation to take advantage of higher producer prices (Stryker *et al*, 1989). Other studies looking at national or global deforestation suggest that trade is only one of the many other factors, such as distorted prices, farmer assistance programs, land tenure systems, and regulations, which lead to deforestation (Reed, 1992; Barbier, 1994).

With specific reference to the economic sustainability of forest firms, under a free trade regime, the improvement of their economic performance sought by improvements in the quantity and quality of production (which however must remain below the sustainably allowable level), and by measures aiming to reduce the costs of operations is of crucial importance. In addition, although still under study and evaluation, current proposals for certification and labeling of forest and forest-based products from sustainably managed forests may exert an influence on forest firms. These proposals have been carried out at an inter-governmental level under the United Nations since 1995, the latest ones submitted in 1997.

This chapter looks at the economic and environmental sustainability prospects of the Greek forest sector in the light of a more liberalized international trade regime. To this end, the sections that follow set the context of sustainable forest development under freer trade, look at the profile of Greek forests, assess the impacts of European Union (EU) forest policies, and conclude by considering the prospects for the sector.

Greek Forests

Greek forests, which extend over a total of 2.5 million Ha, cover some 19% of the country's total land area, and together with the partially forested lands they take up 44% of this area, as seen in Figure 11.1.

From the viewpoint of forest species, 2/3 of them are broadleaved and among these oak and evergreen broadleaved dominate, as seen

Figure 11.1: Forest Distribution by Land Use (%)

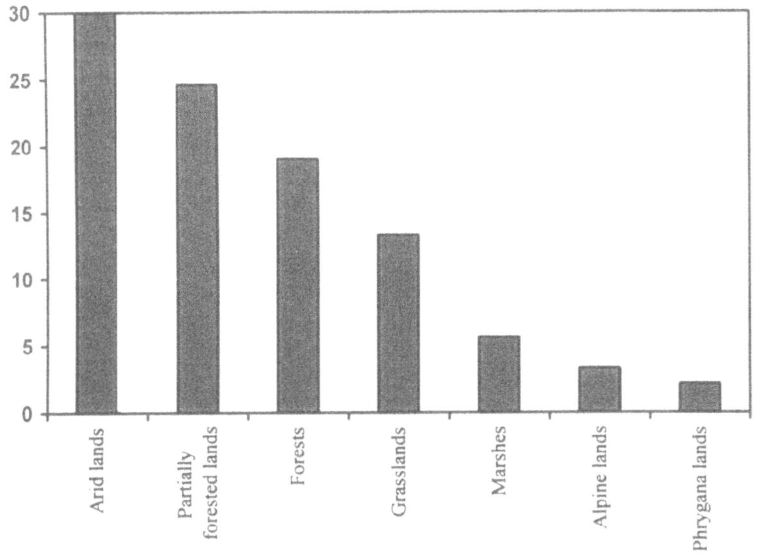

in Table 11.1. Among conifers the *P. Halepensis* and *P. Bruria* forests are the most prevalent in the lower altitudes, often severely struck by fires. The forests of conifers occupy a large percentage of the coastal zone, of high importance for tourism, and from this point of view they establish a very important element for aesthetic evaluation of the natural environment (Eleftheriadis and Tsalikidis, 1989). In the same zone the evergreen and broadleaved are found. The greater part of the oak forests almost and all the evergreen broadleaved are managed as coppice forests. This kind of management is generally considered the least effective for the anti-erosion soil protection and for the performance of the forest's regulatory hydronomic role (ecological functions) since, in the forest practice, it is applied with clear-cut felling. The use of this type of felling in the specific forests of the country is directly connected with the feasibility of the logging works in conjunction with the market prices of produce from these forests which have a low margin of contribution.

From the viewpoint of ownership private forests constitute a relatively small portion (8%) whereas state ownership is the dominant form (65.5%), as seen in Figure 11.2. The ownership of private forests is there for historic reasons, i.e., acquisition by inheritance, donation, etc. Thus, private forest land owners very rarely know the forest from

TABLE 11. 1: Forest size by forest species

Forest species	Forest size %
A. Conifers	
Fir - Spruce (*Abies-Picea*)	13,1
Aleppo pine (*P Halepensis*) - *P. Brutia*	18,9
Black pine (*P. Nigra*)	5,5
Other Conifers	0,9
Total of Conifers	38,4
B. Broadleaves	
Oak (*Quercus*)	29,8
Beech (*Fagus*)	8,7
Other deciduous	4,1
Evergreen	19,0
Total of Broadleaves	61,6
Grand Total	100,0

Source: Ministry of Agriculture, 1986

a scientific aspect and, even more, rarely do they act as managers of their forests. The rule they follow for forest utilization is the annual or multiple-year leasing of the cut (Bazigos, 1987). In this sense, the non-consumptive uses and the ecological functions of the forest, let alone sustainability, have very little importance for private forest land owners because they do not yield any direct income.

The current distribution of forests by form of management is: seedling forests (34.7%), coppice forests(48%), and middle form forests (17.3%). An increase in the rate of the seedling forests producing wood products of large dimensions can be achieved by converting coppice and middle forests into seedling ones. Despite the relative legislative provision in force, this operation has not been launched at satisfactory rates, due to the long time periods required for rotation (>80 years) which involves high investment costs and long payback periods.

Figure 11. 2: Forest Size by Type of Ownership (%)

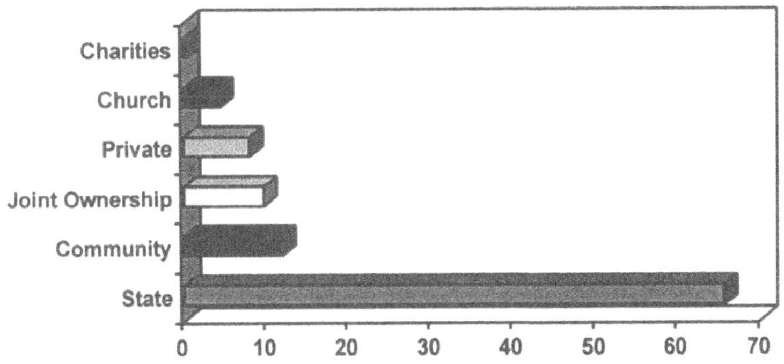

The woodstocking in Greek forests, on the average, is much lower than that of the rest European forests and ranges depending on forest species, ownership, and management form. More woodstocked are the seedling forests, less the middle-form, and even lesser the coppice ones. The average standing stock volume amounts to 74.5 m^3/Ha. Depending on the forest species, woodstocking fluctuation varies from 400 (spruce) to 10 (evergreen broadleaved) m^3/Ha. The average stock volume of conifers amounts to 140, of broadleaved to 65, while of evergreen broadleaves to 10 m^3/Ha. In the state forests the average stock volume amounts to 80 m^3/Ha. In general, the total of forests, from the point of view of the stock volume limit, is below the desired situation (of the sustainable - "normal" stock volume). For this reason, the amount of wood harvested annually (1.2 m^3/year/Ha = 1.45% of stock volume) is smaller than the annual increment of stock volume (2.1 m^3/year/Ha).

<u>Unsustainable uses</u>

While current wood harvesting rates do not appear to threaten the sustainability of Greek forests, other factors such as forest fires and overgrazing do.

Recent research suggests that several factors of varying intensity have been responsible for forest fires in Greece during the last three decades. Among these factors are ecological, economic, cultural, and even political ones (Briassoulis, 1992; Lekakis, 1995). Over 1,500 fires are reported annually to the Forest Service, although the reasons for nearly half of them remain unknown. The most important reasons for forest fires are arsons, pasture burning, stubble burning, cigarettes, dield workers, and trash burning (Stamou, 1995). Although the area burned varies from year to year, it is estimated that 15,000 Ha of forests and 45,000 Ha of forested lands and shrublands are burned annually (Forestry Service, p.c.).

Overgrazing constitutes a serious factor to take into account under a liberalized trade regime. The decline in transhumance in sheep and goat production[1] - a characteristic traditional land-use practice - has had significant impacts on the farmed landscape, by leaving large areas of sub-alpine pasture neglected. Of the pastures (grasslands and forests with grass undergrowth) 58% is suitable for sheep and cattle grazing, while the remaining 42% is better fitted for browsing by goats. Considering the large number of sheep and goats in stable and transhuming flocks, while approximately 85% of all cattle use the lowland pastures, it is obvious that the stocking rate is heavy in the grassland ranges (0.21 sheep/ha) and the shrublands (0.50 ha/goat) (Nastis, 1985). However these figures are even smaller if the actual number of animals is considered (i.e. 0.20 ha/sheep and 0.38 ha/goat). The feeding capacity of Greek pastures appears to be in decline, considering the fact that 25% of the mountainous and semi-mountainous soils have definitely lost their productive capacity, whereas another 40% has been degraded to a dangerous level, due to deforestation and over-exploitation (Alexandris, 1985).

The introduction of new animal breeding structures, the declining transhumation of sheep and goats and the reluctance of breeders to exploit distant alpine grazing and pasture-lands, contribute to the excessive increase of the stocking rate of the rangelands and pastures, a situation which is more evident in the middle altitude areas. These problems do not allow the rational management of pastures and their balanced exploitation. But even under these conditions, pasture production, even in shrublands, helps cover a sizeable part of the nutritional needs of goats, transhuming or not, and constitutes a decisive factor in maintaining the prevailing traditional extensive system of exploiting ruminants (Hatziminaoglou, Zervas, and Boyazoglu, 1995).

There is also a tendency to under-graze or abandon shrublands. Nationally, it is estimated that as much as 40% of all extensive grazing land are under-grazed or have been highly eroded and are now unsuitable for grazing (IEEP and LEI-DLO, 1996). However, not all marginal areas are suffering from abandonment and 'over-exploitation of marginal land is a problem in many regions, particularly as a result of overstocking with sheep and/or goats, largely as a result of the incentive provided by the CAP sheep (and goat) regime' (MAICH, 1997). The ratio of crop to livestock production is 70/30 and this has been blamed mainly on the Common Agricultural policy (CAP). Assuming that this ratio is to change in favor of livestock, as it is also suggested as a desirable policy target (Zioganas, this volume), environmental offences will 'migrate' from the valleys to the rangelands. Rangelands, however, especially the mountainous and semi-mountainous, support important ecological functions. Therefore, they should be exploited in a sustainable manner (Stamou, 1997).

Economic attributes

The economic attributes of Greek forests are expressed in the following indices:

Size of forest per capita of population: 0.25 Ha
Average annual cut: 2.5 million m^3
Number of employed: 35,000 people
Number of work days/year: 2.1 million
Employment in Forestry/Total Employment = 1.19 %
(Employment in Wood Industry/Total Employment=1.25 %)
Added value: 0.1 billion ECU
Contribution to GDP = 1.2%
Imports value/Exports value = 7/1
Production/consumption
- Sawlogs = 24%
- Sawnwood = 50%
- Veneer = 100%
- Chipboard = 100%
- Fiberboard = 14%
- Fuelwood = 100%
- Paper-paperboard = 50%
- Paperpulp= 78%

The size of forests and of the partially forested areas (see Figure 11.1) along with the above lying alpine and the grasslands constitute as a rule large continuous sites which along with the phrygana lands function as renewable natural resources of natural biomass production. From the viewpoint of location, to an overwhelming extent, these forested areas are situated in mountainous and semi-mountainous areas, the vast majority is under public or communal ownership, whereas their continuity is interrupted by very small and intensely scattered private farming lands. The latter, due to the demographic decline of the mountainous areas over the last 30 years, have increasingly converted use from agriculture to pasture. A very recent trend, which however proceeds at a very slow pace, is their placement under the programs of forest plantations on agricultural lands according to the EU Regulation 2080/92.

The distribution of Greek forests by altitude, in conjunction with the prevailing climatic conditions (very dry and relatively long summer, rainfalls in a very short period of the year, often violent) and steep ground slopes provide Greek forests with a particularly important ecological role. This role is especially focused on the protection of soils from the erosion and on its positive contribution to the regulation of the water balance in the soil. Both the above elements are directly connected with the biological sustainability of the natural ecosystems and with the sustainability of the economy.

Consequently, both elements also have their economic and political importance since they are directly connected with: a) the maintenance of productivity of the slopping land in the mountainous and semi-mountainous areas as well as with the maintenance of the base of soil economy of these areas, and, b) the potential of utilizing the runoff waters locally and also (mainly) in low lands where the economy forms, the farming types, and the over-centralization of the population (local and summer visitors) have led for many years to the uncontrollable utilization of ground water deposits. The over-pumping of water as concerns the coastal regions should constitute a special feature to be investigated in conjunction with the role which the suitable management of forests can play in view of the rise in sea level which is expected to begin in the new century.

The multifunctional role of forests and the sensitiveness of forest ecosystems differ inter-regionally. As we are moving away from the North, closer to the tropical zone, forest ecosystem stability decreases, while the ecological functions of forests become more intense (Schutt, 1992). This fact, which depends on the geographic location of the firms and also on the ownership sector and the commitments-constraints which the particularly increased environmental role of its forests enforces on their management and their utilization, makes it possible to differentiate the competitiveness of some forest firms (especially as to the wood products) from the viewpoint of free trade and the need for certification of the sustainable origin of the forest products. In other words, through this international institution it is possible to create comparative advantages/disadvantages for some countries and forest firms.

Greece and broadly the Mediterranean zone and the zones similar to it, depending on the specialized content which the international trend of certification and labeling of products from sustainably managed forests will finally take, may be placed in a more unfavorable position, in comparison with the forms of ownership or other regions with less susceptible forest ecosystems. This fact at the same time poses the problem of the potential of the non public bodies of forest ownership to respond to the environmental constraints which the environmental role of their forests sets on the firm. That is, an increased trend for conversion of private to public forests should be expected in the future, in inverse proportion to the latitude and opposite to the general current trend of privatization of the economy. However, in the developing countries of the tropical zone governments will show inability and reluctance to undergo the above stated environmental constraints.

National and EU Policies and their Impacts on Greek Forests

The European forest policy without being a specialized policy for the forest sector was substantially developed as a by-product of the agricultural and the environmental policy (Loseby, 1996).

The spark for taking measures in respect to forestry was actually given by the willingness and the effort of the European Union to solve the problems of underemployment and low incomes of the less favored areas, the problems arising from the surplus agricultural products, the protection of the natural environment and the reduction of the deficit of trade balance of wood and its products.

The application of the European forest policy which was materialized/or being materialized through the regulations 1609-1610/89 (limitation of agricultural production), 1611 (improving marketing and transformation from cork plantations, 1612/89 improving marketing and transformation of forests products), 3529/86 (improving of degraded forests), 1613/89-1614/89-2158/92 (protection of forests against air pollution and fires), 1615/89 (establishing of European Forest Information System), 867/90 (wood harvesting and transportation measures), 2078/92 (agro-environmental measures), 2080/92 (forestry measures in agriculture), whereas it seems that there is a trend of formulation of a policy for the protection of the environment of the mountainous regions which will have its own impacts on forestry.

From the above regulations, those which refer exclusively to forestry, were launched and applied in Greece to a satisfactory degree whereas those concerning agroforestry were applied with a significant time lag. This lag is due to two main reasons: the delay in awareness of the concerted parties, and the fact that farmers consider an important disadvantage the long-lasting engagement of their farm lands in forest use required for the application of some of these regulations,.

However, it is estimated that the funding of Greek forestry, particularly of state forestry after the application of the system "lease of harvest and trade of cut" of the state forests to the forest cooperatives based on the Presidential Decree 126/86, would have been very poor without the community's contribution through all the previous mentioned measures.

Based on the above, the viewpoint which is formed for Greece is that the regulation 867/90 has functioned in some cases against the natural forest environment because the way in which it was applied has massively displaced the environmentally friendly way of removing wood by animals and at the same time has introduced into the sensitive ecosystems of the mountainous and semi-mountainous forests big and heavy machinery for harvesting and wood transportation resulting in the degradation of the environment. This point of view is directly connected with the above-mentioned system of forest cooperative operation introduced by the Presidential Decree 126/86. The use of heavy machinery aims at making possible the formulation and transport of wood of large dimensions for the wood industry on one hand and at reducing the harvesting cost per cubic meter of harvested wood on the other. It is also estimated that within the frames of the free trade the above regulation in conjunction with the aforesaid system will cause a great deal of talking in Greece in order to apply the certification and labeling of wood from sustainably managed forests.

The use of heavy machinery for wood harvesting and transport along with the environmental degradation which can be caused, comprises a characteristic example of the interregional differentiation and of the creation of comparative advantages/disadvantages described above. It is obvious that the so called environmental degradation from the use of machinery does not occur, or at least does not take place, with the same degree in the northern countries of the temperate zone. There, as already said, the forest ecosystems are characterized by stability and very small biodiversity of species and ecosystems, and the hydrological, water regulatory, and the erosion control role of the forest is not as significant as in the Mediterranean zone.

That is to say, environmental constraints will result in the increase in harvesting costs or generally the reduction of profit margins from the trading of forest products. In addition, they will play a very decisive role in Greece as to whether both non-state and state forestry will be able to remain in the production of wood products to satisfy product needs beyond the local markets. Forest activity might gradually be transferred from consumptive to non consumptive uses and the ecological functions of the Greek forests, since forestry in Greece will continue to pursue a multipurpose function. This role is becoming more of a priority and its significance is expected to increase further and to be directed towards the idea of conservation, sustainable development and the provision of low impact activities. This new trend is evident in all recent studies and plans that have been conducted for the development of mountainous areas and for the introduction of new forms of sustainable development like eco-development and other relative strategies. These actions have identified the need to include local people in much more participatory planning and imply that people should be more aware and involved in the implementation of these plans if the planners want their intentions to be fulfilled and successful. But a consequence of this change might worsen the already negative trade balance in wood, and employment of the population near to forests may either shift out of the wood cutting or decrease, unless appropriate measures are taken in time for a redirection in low impact activities.

The budget of the Forestry Service was, until recently, the driving force of every project concerning Greece's forests and remains to this day very important, although other sources of funding (mainly through the EU) have been created. 23% of the budget of Greek forestry derives from the national budget, 30% from the Program of Public Investments which is itself part of the national budget and 47% from the Central Fund for Agriculture, Animal Breeding, and Forestry. The Service also has an income from selling forest products. These sales account on average for 1/3 of the budget of the Service (Stamou, 1990). Thus it is apparent that the rest of the expenditures are covered by the central government.

The decrease of revenues from state forestry will always put the problem of the funding potential into perspective. Similar consequences are expected to arise from the replacement of clear-cut fellings for environmental reasons wherever they are applied (such as in coppice oak forests) and the establishment in their place by fellings in very small clear-cut felling sites or selective per individual fellings which imply an increased harvesting cost. It would be a challenge for all the countries which practice an environmental rather than a wood production forestry, to develop innovating activities in the methods of wood harvest as well as to apply technology.

Additionally, private forestry does not receive direct assistance for implementing the management plans approved by the Forestry Service. Sometimes owners need to undertake costly projects for the restoration of their areas, or the enhancement of their forests' ability to protect soil and watersheds. Especially when their land has been designated and declared as a forest with an absolute protective function, they are obliged to abandon any economic activity in favor of the management prescribed by the Service, without any compensation. Also, there are no privileges for private owners who seek loans for the improvement of infrastructure and production in their holdings, sometimes also the regulations and the administrative time-lag that an application involves can be restricting factors. There is no concession in the interest rate of the loans available, and sometimes it has been observed that loans granted to agriculture entails much more favorable terms than the ones for forestry. Forest ownership taxation is also problematic, since there are no exceptions or special concessions for owners whose estate lies in very remote and steep lands and thus requires additional expenditures for the management and the collection of the timber. The existing situation is evidently rather disappointing for private forestry and has resulted in a reluctance of the owners to improve stands in their estates and to invest in new machinery and infrastructure.

The need for reviewing the specifications (prescriptions) for management plans and for studies of constructing of every kind dealing with the management and the creation or replacement of infrastructure for the utilization of forests, poses an additional challenge. These specifications, beyond the sustainable cut should take especially into consideration the multifunctional role of forest, which must safeguard and, depending on the situation, upgrade it through the projects, which will be studied and put into action.

Similar adaptations will also undergo the management plans and the projects of utilization of forest rangelands, that according to the forest legislation are forestlands. Here, in addition problems may be caused in the distribution of stock breeding products in the market if it becomes clear that they originate from the utilization of forest resources which suffer from degradation due to overgrazing.

Consequences similar to the above are expected to occur to the wood processing industry which is characterized by low profit margins.

The already strong competition within this industry has pushed many firms to the trade sector. To cope with additional pressures on profit margins stemming from environmental reasons, the wood processing industry needs to proceed to renovation in its production line, products, and marketing (Tanz, 1992). Naturally, the future course of the wood processing industry could be influenced by the developments in agriculture which will designate the availability or not of agricultural lands for the creation of fast growing artificial forests in more leveled soils.

It must be pointed out that in wood manufacturing, international standardization of products has not made a significant progress and therefore the product diversity confers great margins to the larger Greek wood processing industries to cover up needs as much as of the regional as well as of the national and international markets, developing an international marketing.

The above consequences may be the result of an increase in consumer prices of some wood and paper products and an increase of the consumption trends for products of wood substitution which will be cheaper but at the same time much less environment-friendly (i.e. aluminum and other metal materials).

In contrast to the above possibilities, the Greek paper and paper-pulp industries are expected to face intense competition pressures due to their dependence on raw materials from overseas and due to the trends of further enlargement or/and the merger of the already big paper industries of the rich countries. These trends get bigger due to the almost complete standardization of paper and paper-pulp in the world market and the almost complete harmonization of the market. That is, market pressures for the paper production units in Greece are certain and not far away. Certification of products for their origin from sustainably managed projects, according to the 1997 inter-governmental agreement at the UN level, will intensify these pressures. A possibility of confronting this situation is probably the enlargement, after careful study, through mergers with respective units of neighboring countries.

The central organization responsible for the management of Greek forests and the formulation of policy is the Ministry of Agriculture through the General Secretariat of Forests and Natural Environment. Having in mind that the majority of forests as well as forested areas belong to the state, and that even the management plans for the private forests need to be approved by the Forestry service, we can identify the very strong impact of the centrally planned forestry policy in Greece.

The new trend under which policy will be designed at a regional level is becoming a reality, specific responsibilities of the Forestry

Service, as well as some of its staff are passing to the prefectures. This will enable the 54 prefectures of Greece to provide for the forests in their territory within with their development plans. However it has been argued that this practice will bring new problems in forest management particularly in cases where extensive forest blocks expand over two or even three prefectures. Similar problems arise in the case of forest protection against fires. In this case, it has been possible to exclude forest protection from the decentralization of management. However, in the Central Forest Service it seems that there should have already been created an official unit, at beginning for the study and the content, the preconditions, and the consequences of applying into the special conditions of the Greek forestry and wood industry the institution of certification and labeling of products from sustainably managed forests, and furthermore for supervising the application at countrywide level.

In the educational sector seems that the management of the forest's ecological functions and of the non-consumptive uses will increase its share in the total time for educating the trainees.

Conclusions

Greece, a country with strong imports of wood and wood and paper products possesses a forestry sector, which has the profound characteristics of environmental forestry. This strongly environmental dimension is attributed to the specific climate-soil conditions, to the sharply mountainous character of the terrain, and to the relatively short distances from the sea. Among the multifunctional role of forests a special place is occupied by the hydrological, hydronomic, and erosion control function.

Fires and other unsustainable 'uses' of Greek forests create more environmental damage than a hypothetical overexploitation of timber due to freer trade. In fact, international trade and the worldwide interest for the forests is estimated to urge the Greek forestry from consumptive towards non-consumptive uses and ecological functions of forests. This estimation is based on the fact that the aforesaid advances will create comparative advantages in the production of wood and wood products for the countries of the temperate zone which are rich in forest resources and respective disadvantages in the Greek wood production. However, an interesting opportunity is appearing on this point, since if the non-consumptive uses of forestry are to be enhanced in relation to the development of other services, like tourism, the prospects for an improvement of the total forest revenues are high.

The negative impacts for the Greek wood industry may in some extent be ameliorated by improvements and innovations in the production process and international marketing. Similar impacts are expected for the paper industry as well.

It can be concluded that Greek forestry under the free trade perspective and the new developments which drive forest policy decisions and planning in the regional level, will be asked to adapt more quickly innovations and new policy instruments with a view to achieving a better quality of wood products which will fulfill the requirements of certification as products coming from sustainably managed forests and to diversify more its production from wood orientated to services provision. In the latter activities lies the opportunity of forestry development in Greece, since the existing situation of the Greek forests, but also the Greek physical and climatic conditions do not permit the efficient use of these forests for wood production compared to Northern countries where conditions are far more favorable. However the diversity of Greek forests and the existing landscapes can contribute more in the adoption of forest policies directed toward the development of the non-consumptive use of forests.

Note

[1] But, as was previously mentioned, transhumance remains important in many areas due to 'a variety of physical and economic limitations on land use, including the fragility of soils, lack of rain and public ownership of four fifths of all pasture and rough grazing' (IEEP, 1994, p. 24).

References

Alexandris, S. 1985. «Rangelands and Environment,» *Geotechnica,* Proceedings of a Congress on «Rangelands and the Economy of Mountainous Regions,» Thessaloniki (pp. 17-22, in Greek).

Barbier, E. 1994. «The Environmental Effects of Trade in the Forestry Sector,» in OECD, *The Environmental Effects of Trade,* Paris.

Bazigos, P. 1987. «The Private Forests of Greece,» mimeo, Aristotle's University of Thessaloniki, Laboratory of Forest Economics (in Greek).

Briassoulis, H. 1992. «The planning uses of fire: reflections on the Greek experience,» *Journal of Environmental Planning and Management,* 35: 161-173.

Carlowitz, H. C. 1713. «Sylvicultura Economica oder Hausswirthliche Nachricht und naturmaessige Anweisung zur wilden Baum,» *Zucht. J. F.*, Leipzig.

Dafis, S. 1986. *Sylviculture*, Thessaloniki: Japoulis Press (in Greek).

DPIE (Australian Department of Primary Industries and Energy), 1996. «Report of the Chairs.» International Conference on Certification and Labeling of Products from Sustainably Managed Forests, Brisbane.

Eleftheriadis N. and I. Tsalikidis, 1989. *Tourism in the Zone of Coastal Pine Forests*, Technological Institute of Forestry Education, Drama, special edition (in Greek).

Hatziminaoglou, J., N. P. Zervas, and J. Boyazoglu, 1995. «Goat Production systems in the Mediterranean area: the case of Greece,» in A. El Aich, S. Landau, A. Bourbouse, R. Rubino, and P. Morand-Fehr, (eds*) Goat production systems in the Mediterranean*, EAAP Publication No. 71, Wageningen Pers.

Institute for European Environmental Policy (IEEP) and LEI-DLO (Agricultural Economics Research Institute), 1996. *Farming at the Margins. Abandonment or redeployment of agricultural land in Europe*. Report. London, and The Hague, November.

Lekakis, J. N. 1995. «Social and Ecological Correlates of Rural Fires in Greece,» *Journal of Environmental Management,* 43: 41-47.

Loseby, M. 1996. «Attitudes to Forestry in the European Union. A Survey of Approaches by the European Parliament and the European Commission,» in M. Loseby (ed), *Agroforestry and its Impacts on the Environment*, Instituto Nazionale di Economia Agraria, Rome.

MAICH (Mediterranean Agronomic Institute of Chanea), 1997. *Effects of Human Activities on Desertification of Psilorites Mountain*, Department of Environmental and Renewable Resources, Chanea, Crete.

Nastis, A. 1985. «Actual Management of Rangelands,» *Geotechnica*, Proceedings of a Congress on «Rangelands and the Economy of Mountainous Regions,» Thessaloniki (pp. 17-22, in Greek).

Reed, D. (ed), 1992. *Structural Adjustment and the Environment*, Boulder, Colorado: Westview Press.

Repetto, R. 1989. «Economic Incentives for Sustainable Production,» in G. Schramm and J. Watford (eds), *Environmental Management and Economic Development*, Baltimore, MD: The Johns Hopkins University Press.

Schutt, P. 1992. «Sustainability of Forests: An Attempt at a Biological Interpretation,» paper presented the Conference «Environmental Ethics, Sustainability, Competition,and Forestry,» University of British Columbia, 23-24 October.

Stamou, N. 1990. «Forests and Mountain Economy, Problems and Prospects,» *Scientific Annals of the Forestry Department*, University of Thessaloniki (pp. 477-486, in Greek).

Stamou, N. 1995. «Analysis of the Forest Fires Caused by Arson in Greece,» Greek Forest Fires Research project, intermediate report to the Ministry of

Agriculture, the Secretariat General for Research & Technology, and the NATO-SFS (Science for Stability), Aristotle University of Thessaloniki, April.

Stamou, N. 1997. «The Meaning of Sustained Yield in Use of Forest Rangelands,» proceedings of «Sustainable Utilization of Range-lands,» 1st Pan-Hellenic Congress of Hellenic Range and Pastures Society, 6-8 November, Drama, Greece (in Greek).

Stryker, J. D. *et al*, 1989. «Linkages between Policy Reform and Natural Resources Management in Sub-Saharan Africa,» unpublished paper, Tufts University, Fletcher School and Associates for International Development.

Tanz, J. 1992. «Sustainable Development, Sustained Yield, and the Meaning of Life,» paper presented at the Conference «Environmental Ethics, Sustainability, Competition, and Forestry,» University of British Columbia, 23-24 October.

Zioganas, C. M. (this volume), «The Organization of Primary Agricultural production in Greece: Evolution and New Challenges for Re-Orientation.»

Part III
The Future

Chapter 12

Greek Agriculture at the Crossroads: Competition and Sustainability

George A. Daoutopoulos and *Myrto Pyrovetsi*

Introduction

Greek agriculture has experienced very rapid changes in the post World War II period (Myrick and Witucki, 1971; Sanders, 1962; Shaw, 1969; Weintraub and Shapira, 1975). Changes were very rapid, considering the catastrophes of World War II and the civil war that followed. The infrastructure and the productive capacity of present day Greek agriculture have been substantially transformed compared to pre-war structures.

The development path adopted by Greek agriculture was similar to the one followed by other western countries; that is, increasing inputs accompanied by major public investments in infrastructure (reclaiming new lands through drainage, irrigation projects, etc.). The use of chemical fertilizers, insecticides, pesticides, new high-yielding varieties, more machinery, and new farming practices all expanded tremendously.

The post-war transformation of Greek agriculture led to a dramatic increase in farm income. This improvement was accomplished without any increase in farm size. Farm size according to the last three agricultural censuses did not change much (see also Zioganas, this volume). During the period 1961-1991 it ranged between 3.6 to 3.9 ha. From 1950 to present, the average land farmed has been about four hectares, split into six parcels. Therefore one must conclude that increases in farm income have resulted from the better productivity of the farm sector. The parallel development of the overall Greek economy was primarily based on foreign exchange earned from agricultural exports. Until 1972, agricultural exports made up the majority of the total exports.

Accession to the European Union

Accession of Greece to the EU was welcomed for the new opportunities to be opened for all sectors of the economy. Agriculture was considered as the sector to benefit most from the new markets because it had the capacity to compete successfully with the similar sectors of the other member states.

Unfortunately the results obtained so far have not lived up to the expectations (Demoussis and Sarris, 1988; Georgakopoulos, 1988; Zioganas, 1993). The original surplus in the agricultural trade balance with EU has rapidly eroded and is presently in a deficit status. The experience of Greek agriculture, so far, is one of lost opportunities. EU financial supports, instead of being used to improve infrastructure have been distributed as subsidies to farmers. The farmers, in turn, have not invested their surplus income in their farm operation. As a consequence, most Greek meat products do not comply with EU standards. Animals are slaughtered in conditions only slightly different from those existing centuries ago. Regional agricultural markets have not been completed so far, leaving marketing of agricultural produce to the hands of occasional traders, with no facilities or brand names.

At the same time, the Common Agricultural Policy (CAP) did not overcome the inequalities between Northern and Southern countries, or lead to a balanced development of all agricultural sectors. The statistical evidence is overwhelming. Eighty percent of the total financial supports for the agricultural sector in the 15 member states are received by 20 percent of the agricultural holdings. Excess capacity of some production sectors is getting worse and runs contradictory to the basic principles of agricultural sustainability. Forty percent of FEOGA (European Agricultural Guidance and Guarantee Fund) expenditures is directed to expanding monocultures (durum wheat, cotton, peach trees).

In the rest of the chapter we comment on aspects of the Greek farming sector which are crucial to its future prosperity.

Co-op Underdevelopment and the Farm Movement

In the past, the underdevelopment of Greek cooperatives was blamed on outdated legal provisions. Repeated revisions occurring in the last two decades have rendered this argument as completely false. Despite three reform acts passed by the parliament there was no substantial development of the Greek cooperatives (Daoutopoulos, 1996). On the contrary, there was a substantial decline in the cooperatives. Cooperative factories have closed all over the country, while those that remain in business are in a very bad financial situation. The share of cooperatives in the overall marketing of agricultural produce has dropped dramatically.

Unfortunately, there is little grassroots understanding of the importance of cooperatives. Because of this lack of understanding all development efforts have been 'top down' processes. Cooperatives have been praised in the past as a means to secure survival of small farm operations that are unable to compete with large-scale operations. In this aspect, any means to preserve small-scale farming has a very important consequence for sustainability.

Drastic measures are needed. In one case European Parliament Member Mr. Papagianakis (1997) proposed the «dissolution of all cooperatives in one night and their complete reorganization on the next day». While we agree on the immediate closure of presently operating cooperatives we disagree on their reorganization even under the principles he suggested (voluntary, higher member contribution of capital, and highly qualified personnel under a new organization scheme). Instead we propose the establishment of 'Popular Companies' in place of cooperatives. The new companies will inherit the assets of the former cooperatives while the members will contribute substantial new capital to secure a healthy financial situation. Popular companies, while having all the positive characteristics of the cooperatives and capital stock enterprises, do not retain the negative characteristics of both types. For example, in coastal navigation, popular companies that were formed in the late sixties have succeeded despite severe competition. The initiative can benefit from the incentives provided to farmers who form groups of producers according to the EU regulation 2200/96.

Unrest in the farm sector has become evident during the last four years. Even so, the magnitude and the severity of the episodes have surprised the Greek public. The unrest was an inevitable outcome of the farm crisis during the last five years. In November 1993, after a one day meeting with farmers in a rural town in Northern Greece we wrote in a commentary (Daoutopoulos, 1994):

> "...here comes the issue of marketing of agricultural produce which in our country continues to follow the provisions of the last century. If marketing is not reorganized drastically following the European standards one does not have to be a prophet to foresee that severe rural unrest is expected not only in *Nevrokopi* but in many other regions of the country that face nevertheless the same problems".

The official explanation that blames either opposition parties or large landowners or a mimic syndrome (since every group in the country blocks national traffic why not farmers too?) is not true. Similar official explanations were provided in every rural unrest episode during the last four years and aim at calming the Greek public. In the 1993 unrest with tobacco growers in *Pieria* (at the climax of the episodes a farmer burned himself), the Minister of Agriculture provided the same explanations used by the present Minister. We hope he won't repeat them in the next series of episodes that, in our opinion, will involve more people and more actions of violence since the political leadership has not understood the real forces behind these unrest and does not seek to cure the factors responsible. For example, what is expected to happen if subsidies are severely restricted? A county agricultural official gave the following description, «the money we distribute in our Prefecture is more than 15 billion drachmas (54.5 million U.S. dollars). The entire Public Investment Program of the region is ten times less. What would happen to our farmers in case of drastic cuts? The phenomena of withdrawal evidenced by chronic narcotic users will be nothing compared to those expressed by our farmers!»

Agricultural Extensions

Following the rural unrest of 1995 and 1996, the official leadership of the Greek ministry of agriculture declared that one of the reasons that brought farmers into the streets was the lack of adequate information. Of course lack of communication between the Ministry and the farming population was a problem reported years ago (Ananikas and Daoutopoulos, 1980) to which no corrective action was taken. If, as a result of the rural movement, the Ministry takes the necessary actions, it has to be considered positively.

Although our society is gradually heading to a new structure where information becomes the primary commodity (Toffler, 1970), the Ministry of Agriculture, during the last twenty years, has followed the opposite path. The Directorate of Agricultural Extensions, founded in 1951, and to which much of the postwar Greek agricultural development is credited, is at a stage of dissolution. The service has abolished most of its educational and informational component and has assumed bureaucratic functions by checking the qualifications of farmers applying for subsidies.

At the same time, billions of drachmas are spent each year for unnecessary training. A much smaller amount would have been enough to reorganize the Agricultural Training Centers as the educational component of the regional agricultural development plan.

For those who assume that farmers will learn eventually through other channels of information –it seems that there are many proponents of this view among the leadership of the Ministry of Agriculture - empirical research has provided some alarming evidence (Pyrovetsi and Daoutopoulos, 1997, 1997a). In wetlands - fragile areas with great biological importance - the ignorance of farmers has had devastating impacts on the environment, even worse than the impacts from the ignorance of farmers in the plains. Without an efficient educational component, the effort to disseminate sustainable farming practices will not bring substantial results.

Agricultural Research and the System of Statistics

Agricultural research, as an equally important factor in agricultural development, has a very crucial role to play in making Greek agriculture sustainable. In Greek agricultural research we have both brilliant and meager results. In grains and cotton, for example, indigenous research has developed varieties better adapted to local conditions. Unfortunately, research on the economic and social aspects of farming in Greece is lagging behind. A group of three scientists has assumed this difficult and very important task.

The National Center of Social Research, with an important contribution in the study of rural Greece, is recently recovering after a long period of severe budget cuts and restrictions in their research programs. In addition, academic research in the major Universities through EU funding and cooperation with the European scientific community, tries to cover the missing pebbles of knowledge in the mosaic of the Greek rural society. Unfortunately. too many pebbles are missing to complete the real picture. As a result, we try through limited resources to draw the complete picture that frequently does not match reality.

This limited and fragmented research system is combined with an incomplete statistical system to make understanding of the rural world even more difficult. Improvements carried out during the last decade have not yet produced visible results.

The Reorganization of Production

Looking back at the literature on Greek agriculture during the last 30 years or at the informational/educational programs carried out by the agricultural training centers throughout the country, one would discover that the need to reorganize the production is an old theme. Researchers in the rural scene frequently hear farmers saying the word 'problematic' to be followed every time by a different crop. ''Barley tobacco is problematic', 'Kiwis are problematic,' 'peaches are problematic' and so on to be accompanied by an agonized question, 'What crop should we plant this year, agricultural officer?'

The policy to promote 'dynamic crops' is not a solution for every farmer and has failed to solve the problems of small acreage and inadequate income. A holistic approach is needed taking into account the specific conditions prevailing in each area that are totally ignored by the uniform measures of the CAP. An overall policy to promote the development of rural areas should take into consideration site-specific conditions, accompanied by measures and initiatives having a sustainable character. In 1995 a two week field trip in the island of *Lesvos,* in the Aegean sea, brought the following notes (Daoutopoulos, 1995):

> Instead of promoting the sheep and goat industry, in an island with limited and degraded pastures, little rainfall and limited feed, an opposite policy is required. Offer subsidies to shepherds who will limit their stock and through training, educate them in ways to manage pastures in a sustainable way (not overgraze, restore plants, conserve water resources, etc.). The island of *Lemnos* exports dairy and meat products to neighboring islands. Presently, the island is completely deprived of forests. The enumeration of trees won't give more than 500. The island has 6.000 abandoned wells with brackish water, highly degraded pastures, numerous abandoned sheep barns, recently studied by a government official. Furthermore, the visitor of the island sees emaciated cows grazing in harvested grain fields and farmers irrigating with artificial rain alfalfa fields The theater of farm madness in its supreme presentation.

The Problem of Risk

Risk management is certainly an important area for future farm policy concerns. In a free-market, there are several risk-increasing factors for South European farmers. Among these are reformed CAP provisions, the industrialization of agriculture, and the reorganization and globalization of international markets. Additional risk factors include changes in technology, environmental and health problems, and the transfer through international transportation of various plant and animal diseases. Risk comes in various interrelated forms: price, revenue, and income risk.

The 1992 CAP reform and the reform proposed by «Agenda 2000» have switched the focus of farm policy from price support to income support. Although this may imply lower producer prices, their level will more or less remain stable over the next few years. However, the problem of price and revenue risks will come up more intensely after those years are over. Income payments under the CAP offer substantial support but they do not provide direct protection against income risk.

In the course of adjusting to the new business environment, large farmers with better organization and resources may be able to deploy risk management strategies such as crop and livestock insurance and marketing, but the medium and especially small farms will indeed face a problem. The sole insurance programs currently provided in agriculture by the government and a company controlled by the Agricultural Bank concern catastrophic insurance. Naturally, these programs cannot cover the risks from operating in a free market world.

It may, therefore, be inescapable to eventually follow models of risk management already applied across the Atlantic. The US Department of Agriculture and private insurance companies have developed three major programs, Income Protection (IP), Crop Revenue Coverage (CRC), and Revenue Assurance (RA), which are there to assist the farmer's own risk management options (Harwood, Goble, and Perry, 1997).[1] There are two problems with such programs that are relevant to competitiveness and sustainability. Participation by farmers involves paying premiums that increase production costs, so small farmers may be excluded *ipso facto*. And, the crops cultivated by large farms that are covered by the programs are usually soil erosive and chemical intensive, such as cotton and corn. Therefore, designing government crop insurance programs involves taking into account not only the economic sustainability of small farms, which are spread all over Greece, but also ecological sustainability which is certainly not guaranteed by the continued production of monocultures. Finally, assisting farmers to become knowledgeable about the risks inherent in their investment decisions and the need for risk management strategies is of the outmost importance.

The Road to Sustainability

Proponents of freer trade assume that with open markets and through competition, every region or country will eventually produce those agricultural products to which it has a competitive advantage. Unfortunately, this basic economic principle often runs contradictory to ecological principles and sustainability. It is not ecologically wise to grow the same crop over large areas, simply because the specific region has a competitive advantage to this particular crop.

On the other hand, it is well known that there are no perfect markets capable of providing a fair price to the factors of production. Therefore, markets should not be allowed to function on their own and the rhetoric of free markets should be downplayed, as markets do not provide the solution to production problems, especially if ecological aspects are considered. Instead, agricultural markets should be supervised and well organized to provide a fair price to producers who, as in the case of Greece, are usually the losers.

A retrospective analysis of Greek agriculture will reveal that until the1950's it retained many sustainable characteristics, such as: very limited use of chemicals; use of indigenous genetic material of high biodiversity; caused very little pollution of waters and of the environment in general; lack of monocultures, as a system of multiple crops was followed; incorporation of animal and crop production in a mixed type of farms benefiting from the exchange of byproducts; complete recycle of produce and lack of garbage; and, small energy inputs

At the same time, local culture as evolved through centuries of interaction between humans and their environment had a remarkable sustainable character that deserves to be mentioned. In Lake Prespa, for example, farmers were harvesting the reeds for animal food and building material. This practice was very beneficial for the lake, since it was removing every year large quantities of nutrients (Pyrovetsi, 1984). Buildings (houses and farm buildings) where constructed with mud and straw if local stones were not used. When those buildings were demolished, the material was recycled to the natural environment. Water from the lake was used directly by humans and animals until the 1970's. Presently, reeds are not harvested and nutrients in the lake have increased, through the chemical fertilizers used in the intensive farming practices. As a result, the lake is eutrophic and even animals do not drink its water.

Local people, through experience gathered for centuries, have invented cultural elements to support a symbiotic relationship with the natural environment. Modern agriculture that followed, increased productivity but at the same time impacted heavily on natural ecosystems and on their ability to supply indefinitely valuable products for the survival of human beings.

After five decades of intensive efforts to increase inputs in Greek agriculture, the message has to change. The road to sustainability should become the future route for Greek agriculture, embraced by all actors (producers, scientists, ministry of agriculture and the academic community). It is said that in the coming century, developed agriculture will be considered the one that has resolved successfully the problems associated with the conservation of the natural environment. Greek agriculture has some comparative advantages, which, in an era of high input agriculture, are considered as disadvantages, such as:

a) The variety of microclimatic conditions, soil conditions and the cultivation of different crops in small parcels contribute substantially to diversity, a valuable characteristic in sustainable agriculture,

b) Although chemical inputs have increased, they are still far below the corresponding levels in other European countries,

c) Fragmentation blocks prevalence of monocultures over large areas and while a disadvantage in modern agriculture becomes an advantage in sustainable agriculture,

d) Proximity of farm-land with forests, water bodies and small plains (where intensive agriculture is practiced) offers a variety of ecosystems and habitats for wildlife that is beneficial to agricultural production (crop pollinators, beneficial insects, predators and parasites of pests),

The road to sustainability is not easy. All participants (producers, scientists, leadership, and local society) must be involved in the process. In this process, a number of factors should be given special emphasis, such as:

a) Better training of producers, since sustainable agriculture requires qualified personnel than conventional agriculture. Agricultural Extensions and training centers should resume this task,

b) Drastic changes in the marketing of agricultural Produce that today resembles markets of the Orient. Agricultural products should lose their generic character and become eponymous and marketed after careful quality control and packaging.

c) Dissolution of the cooperatives and creation of new forms of associations, based on a wide distribution of shares among local producers,

d) Formation of a land policy to preserve valuable and non-renewable fertile land resources (Daoutopoulos, 1980; Daoutopoulos and Pyrovetsi, 1995),

e) Reorganization of the Agricultural administration, especially of that in the Ministry of Agriculture and formation of a new policy that will assist in the gradual shift to a sustainable form of agriculture,

f) Promotion of agricultural research in all production sectors,

g) Preparation of integrated development programs for every region, tailored to the specific characteristics.

Sustainable agriculture and sustainable development of rural areas require actions and measures that will allow local people to respond to the changing environment, to heal the social network and become again capable of producing its own culture.

Note

[1] IP offers an annual «revenue guarantee» based on producer planted area, expected yield, and sign-up time futures price for harvest-time delivery. If the harvest-time price multiplied by the actual yield in that year falls below the guarantee, the producer receives a compensatory amount of money. CRC is similar to IP but it contains an additional element. If a farmer's yield falls short and the projected price is lower that the actual harvest price, he is compensated for the yield loss at the harvest price in order to allow him to purchase 'replacement' crop quantities in the market. RA is also similar to IP. However, whereas IP allows a farmer to insure only at the farm level as an enterprise, RA has a unique system of premium discounts depending on the farm structure.

References

Ananikas, L., and G. Daoutopoulos. 1980. «The Human Factor and its Importance in Agricultural Extensions and Rural Development,» *Geotechnica*, (3):2-24 (in Greek).

Daoutopoulos, G. 1980. «The importance of preserving farmland for the Greek Economy,» Thessaloniki, Society of Economists, Special series No 7. Also published in *Oikonomikos Tachydromos*, (35-36) August-September 1981, and in the Scientific Publications of the Agricultural Bank of Greece (Epistimonikes Ekdoseis ATE), 1980 (14): 84-113.

Daoutopoulos, G. 1994. «The contribution of Agricultural Scientists in the Prosperity of Rural Greece: A Recent Experience,» *Geotechnical Review*, 59: 59-61 (in Greek).

Daoutopoulos, G. 1996. «Humans and Natural Environment,» *Geotechnical Review*, 70: 73-74 (in Greek).

Daoutopoulos. G. 1995. «Agricultural Cooperation in Greece: The Case of a Village Cooperative Association,» *Journal of Rural Cooperation*, 24(2): 91-104.

Daoutopoulos, G., and M. Pyrovetsi, 1995. «The Need of Conserving Farmland to Meet the Demand for Food,» *MEDIT*, 3:18-25

Demoussis, M. and A. Sarris, 1988. «Greek Experience Under the CAP: Lessons and Outlook,» *European Review of Agricultural Economics*, 15(1): 89-107.

Harwood, L., K. Goble, and J. Perry, 1997. «Alternatives for Producer Risk Management,» in USDA, *Agricultural Outlook Forum, 97*, proceedings , Washington, DC, February 24-25.

Georgakopoulos, T. 1988. «The Impact of Accession on Agricultural Incomes in Greece,» *European Review of Agricultural Economics*, 15(1): 79-88.

Myrick, D. C., and L. A. Witucki. 1971. *How Greece Developed Its Agriculture*. Washington, D.C.: USDA Foreign Agricultural Economic Report No 67.

Papagiannakis, M. 1997. «Greek Agriculture and its European Future,» Proceedings, Thessaloniki, PASEGES (in Greek).

Pyrovetsi, M. 1984 «Ecodevelopment in Prespa National Park, Greece,» Ph.D. Dissertation, Michigan State University, East Lansing, Michigan, USA.

Pyrovetsi, M. D. and G. Daoutopoulos, 1997. «Contrasts in conservation attitudes and agricultural practices between farmers operating in wetlands and a plain in Macedonia, Greece,» *Environmental Conservation*, 24(1): 76-82.

Pyrovetsi, M., and Daoutopoulos, G. 1997a. «Farmers' Needs For Nature Conservation Education in Greece,» *Journal of Environmental Management* (in press).

Sanders, I. 1962. *Rainbow in the Rock: The People of Rural Greece*, Cambridge, Mass: Harvard University Press.

Shaw, L. H. 1969. *Postwar Growth in Greek Agricultural Production*, special studies series No 2. Athens, Greece: Center of Planning and Economic Research.

Toffler, A. 1970. *Future Shock*. New York: Random House.

Weintraub D., and M. Shapira 1975. *Rural Reconstruction in Greece: Differential Social Prerequisites and Achievements During the Development Process*, Beverly Hills, California: Sage Publications.

Zioganas, C. M. 1993. «Structure and development of Greek agriculture before and after accession to the E.C.,» *Quarterly Journal of International Agriculture*, 32(2): 47-159.

Zioganas, C. M. (this volume) «The Organization of Primary Agricultural Production in Greece: Evolution and New Challenges for Re-Orientation.»

Chapter 13

From Common Agricultural Policy (CAP) to Common Agricultural Planning and Management (CAPLAM)

Stavros M. Theofanides

Introduction

It is well known that the EU faces some serious problems in agriculture. These problems are both short-term and long-term and include the following: (i) persistent surpluses in some sectors; (ii) rising financial costs in supporting the Common Agricultural Policy (CAP); (iii) failure to achieve structural changes and to correct the increasing imbalance between production and demand; (iv) failure to support regional balances in the farming areas of the EU, and an equitable distribution of benefits and costs among the member countries; (v) ignorance of high quality considerations of agricultural products in determining the CAP and the failure to exploit the comparative advantage of the EU in some agricultural products in international markets; (vi) frequent unrest and mobilization of European farmers demanding more support for agriculture; (vii) the weak stand of the EU on the problem of world famine; (viii) persistent uncertainty in agriculture mainly due to the lack of long-term planned action in the agricultural sector; (ix) the confusion of politics and economics in determining the CAP; (x) ignorance of the need for environmental sustainability in framing the CAP; (xi) failure to take into account the preservation of the cultural heritage embodied in the agricultural sector; and, (xii) ignorance of the farming community's particular status as a social group with an idiosyncratic way of life. Most of these problems arise from the fact that the agricultural affairs in the EU are arranged and managed through the 'method of policy.' A more efficient strategy to follow would be the 'method of planning and management' at micro, sectoral, and macro levels.

This chapter supports the thesis that the CAP should be replaced by CAPLAM; a more efficient instrument to tackle the «agricultural problem» of Europe and in particular the problems (i)-(xii) mentioned above. However, the chapter also suggests that the proposed 'method of agricultural planning' should be market-oriented rather than centrally controlled. In this context, planning would define 'desired targets' while market forces would function to allocate resources rationally in order to achieve the planned targets. In comparison to the political maneuvering and spasmodic measures employed by the existing CAP, the long term strategy of planning in a market-oriented world seems to be a superior instrument in the efficient management of agricultural affairs The agricultural policy of the EU must be conducted with a three-dimensional architecture. First, the purely economic dimension based on market-oriented planning objectives and targets. Second, the environmental sustainability and regionality dimension. Third, the social status-cultural heritage dimension.

The above three premises of agricultural policy are closely interrelated and must be considered together. Defining or planning only one dimension would not be adequate. Agricultural activity (not only production) has an important influence on environmental sustainability. Also, farming is performed in a given social context which has deep cultural roots related to agricultural life. Therefore, it is necessary to understand European agricultural policy in totality, that is taking into consideration all economic, environmental, social and cultural dimensions. The definition of European agriculture on the basis of solely economic considerations is a great mistake since it ignores the fundamental nature of agriculture as an interrelated economic, nature-based, environmental, and socio-cultural activity.

Major Mistakes of the CAP

The first basic mistake of the CAP is its ignorance of the market demand for agricultural products. It would seem that the CAP was designed in an economic vacuum, that is, without the taking market forces into consideration and especially ignoring the fact that the supply of agricultural products is designed to satisfy a certain level of demand. The general functional form of demand for a particular commodity is expressed by the well-known function,

$$Q_d = f(P_c, P_1 \dots P_n, Y, N, T_r, D) \qquad [1]$$

where:

Q_d = the quantity demanded during a given period,
P_c = is the price of the commodity,
$P_1 \ldots P_n$ = the prices of other commodities,
Y = the average per capita disposable income,
N = the population,
T_r = the preferences and tastes of the population for a given product (which is not very strong for agricultural products), and,
D = the distribution of income within the population.

The generalized demand function expressed in equation [1] presents the six variables that influence the demand for a given commodity or agricultural products in general. Equation [1] «defines» how much of a product (all agricultural products) will the consumers buy from the market.

The strategic variables in equation [1] for the demand of any agricultural commodity concerning the CAP include, the per capita disposable income and «the preferences and tastes» of the 370.000.000 people of the European Union (EU). Moreover, total demand for agricultural products must cover two other «exogenous» components: + 2 per cent of the total production allocated to agricultural reserves for unforeseen contingencies, and + 1-2 per cent (or more) for international aid to third world countries.

The second mistake is the insensitivity to differing quality of various agricultural products. Nearly all agricultural products produced in the Mediterranean countries are of superior quality due to ecological and environmental differences compared to the Central and the Northern regions of Europe (Germany, Belgium, Netherlands, G. Britain, Denmark, Sweden, Luxembourg, Finland, Ireland, Austria). In chemical terms, the nutritional value of a kg of Greek apples or grapes is far superior to the same products produced in Germany, Denmark or Sweden. The same is also true for tomatoes, cucumbers, lettuces, and other goods. If agricultural prices were to be formulated according to the above mentioned quality differences and chemical standards, the prices of Mediterranean products should increase by a factor of between 2 and 3. In the long-term Europeans must accept that the quality differences of agricultural products varies across Europe due to differences in the climate. This factor is imperative in designing the future CAP on a rational basis. Europeans must buy their pulp paper from Sweden and Finland but it is healthier to eat apples and grapes or lettuces from Greece and Italy. Thus, the second most important mistake of CAP is that it is totally ignoring the chemical composition of the origin of agricultural products.

The third mistake of the CAP has been to ignore the concept of a *Total Agricultural Income (TAY)*. A new and more pragmatic definition of the main economic variable of the farming unit is required . The orthodox definition of the gross agricultural income *(AY)* includes the proceeds of the sales of agricultural products, that is:

$$AY = Q \bullet Pr \qquad [2]$$

where:

Q = the quantity of agricultural products produced by a given farming unit, and,

Pr = the average prices offered to the producers of Q during the sales of products.

This is a very narrow definition of the gross agricultural income; in fact, it underestimates the agricultural income enjoyed by farming units de facto, since it does not include income from some other sources. Farmers today have incomes from many sources and the CAP does not take into account the broad definition of agricultural income in this context. In the following definitional equation [3] we try to expand the definition of *TAY*, to include all potential sources of income generated by farmers, taking into consideration also the «psychic income» which is enjoyed by the farming population due to the cleaner environment in which they live compared with the urban population.

The following equation gives the main ingredients of the total agricultural income:

$$TAY = AY + RY + MY + TY + SY + CY + TrY + DY + OY + [PsY] \qquad [3]$$

where:

TAY = total agricultural income,

AY = income from sales of agricultural products: $Q \bullet Pr = AY$ (where: Q = Quantity of Products, Pr = Prices of prod ucts),

RY = income from the processing of agricultural products,

MY = income from marketing agricultural products,

TY = income from agro-tourism

SY = income from supporting or complementary activities,

CY = income from primary cultural activities,

TrY = income from transfer payments (subsidies etc.),

DY = imputed income from the non-distribution costs of the products for self-consumption (on-farm consumption),

OY = other income or incomes from other sources,

PsY = imputed psychic income due to the «Arcadian life» next to nature.

It is well-known that some of the above sources of additional incomes may sometimes offer higher components than the direct income from the sales of agricultural products. It is interesting to note that in case that the farmers undertake the marketing of their own products they could enjoy 40-65 % higher incomes than the average from the sales of their products. Certainly the valid relation is:

$$TAY > AY \qquad [4]$$

even when we do not include in our definition of *TAY* the imputed psychic income. «Arcadian life» and «Arcadian income» is the income enjoyed because of the living of farmers in the rural paradise as compared to the urban environment. Arcadia is a famous agricultural region of ancient Greece considered idyllic, pastoral, simple or untroubled. It is a great mistake that the CAP is mainly based on the concept of gross agricultural income (AY) and ignores the broader definition of the farm income *(TAY)* as given above in [3]. In fact, a more accurate definition must distinguish between 'realized' and 'imputed' income. However, both these incomes contribute to the welfare of the farmers and, therefore, *TAY* as defined above is a more representative variable of the income enjoyed by the farmers.

The fourth mistake of the CAP has been to ignore the importance of *Total Quality Management (TQM)* in the operation of farms. Management plays a crucial role in agriculture both at the sectoral and the farm level. It includes strategic functions such as the planning, organization, staffing, direction, coordination, communication, budgeting and control which are necessary in the operation of any entrepreneurial activity as is the farming unit. It is common place that the success of agriculture and any policy designed to support it as, for example, the CAP depends on the techno-economic performance of the farming units. Any policy objectives are realized at the farm level on the basis of managerial performance. Efficient management techniques produce miracles in any activity including farming. Production efficiency, cost minimization, high investment profitability, income stability, the use of modern farming techniques and other objectives of the farm firm are achieved through high performance total quality management. Regretfully, the CAP ignores the managerial function at the farm level or gives too little emphasis to such an important factor for the progress of agriculture. The very essence of our proposed strategy is based on implementing management both at the sectoral and the farming unit levels. The strategy encourages management education and training, the creation of «centers of agricultural excellence» and other schemes promoting the managerial capability of farmers.

The fifth mistake of the CAP has been to neglect the importance of the *Cultural Heritage of Agriculture (CHA).* Culture includes the basic values and customs which characterize the patterns of our everyday life and determined our behavior. Culture is a vital and important aspect of our lives and plays a crucial role in some social groups such as the agricultural population. The farming society lives in different cultural geographical areas known in cultural science as *Kulturprovinz* (German: «culture province»). These areas are living museums of the primitive and post-primitive life of humans and have a special interest for the modern industrialized and mechanized societies. Cultural practices from certain artistic to ritualistic patterns which are manifested in work customs, dressing, entertainment habits and other functions are very interesting not only to anthropologists and sociologists (like Alfred Louis Kroeber and Ruth Benedict) but also to policy-makers. It is a pity that the CAP ignores the cultural heritage of agriculture. Any comprehensive policy on agriculture must take into consideration the preservation of the particular cultural identity of this vital sector of the European economy. The strategy could be the following: The cultural heritage of agriculture could be linked with agro-tourism thus offering an important source of income to the farmers as is explained in equation [3] on total agricultural income.

Criteria for a Better Strategy

Priority to the Ecology-based biochemical quality of products: Agricultural produce is directly related to human life and health through nutrition. This is the most vital (literally speaking) aspect of agriculture as an indispensable economic activity. It is necessary to re-orient European agriculture towards those ecological regions which through sun, soil and other climatic characteristics secure agricultural products (both in the arable sector and the livestock sector) of the highest chemical quality. This is the most important factor in our considerations for the formation of long-term agricultural policy. All considerations about free trade, competitiveness, support costs, and trade interests (topics which preoccupy many experts) are rather insignificant *vis-a-vis* the nutritional-quality aspect of the agricultural products.

Mediterranize EU agriculture: Cleopatra, the famous attractive and clever woman of antiquity though living in Egypt used to order her honey, wine and figs from the island of Thasos in Macedonia, Greece. It would benefit the overall health of Europeans (and others) to get their agricultural products from the Mediterranean region and to try to take the pollution to the north or central Europe, inhibiting manufacturing which may poison the agricultural products. The strategy of giving a Mediterranean quality to selected agricultural products will be beneficial to *all* European consumers.

Secure both an economic and financial «balance» in the terms of trade between agricultural and manufacturing products: There is a widespread belief that the long-term terms of trade between agricultural and manufacturing products are against agricultural products. This half-myth is not true when we «deflate» the comparison of the two categories of products in quality terms. It is considered that if we improve the ecological-chemical quality of the agricultural products by restricting or guiding their production in the Mediterranean region then the superior quality agricultural products (given that their supply will be limited away from central and north Europe) will improve their terms of trade *vis-a-vis* the manufacturing products.

Promote the regional specialization of all agricultural products in Europe according to the ecological-quality criterion: this will secure the highest nutritional intake from the consumption of the regionalized agricultural products for the European consumer and will eventually lead to higher living standards and individual health for the whole European population.

Apply the Linear Programming (LP) philosophy and techniques: Policy makers seeking the best policy for European agriculture must be inspired and directed by the philosophy and the techniques of LP. Generally speaking, this is a mathematical tool for solving maximization and minimization problems when constraints restrict the course of action. There are maximization aspects (such as the farm income of agricultural households) and minimization ones (such as the cost of production from various input allocations) and, of course, budget constraints due to the budgeting limits of the European Agricultural Guidance and Guarantee Fund, FEOGA.

LP is very useful in solving constrained maximization and minimization problems for a given farming unit. Specific applications of the technique in computerized case-studies or models must be done in selected agricultural areas on the basis of pre-determined choices according to the particular ecological-climatic conditions prevailing in the selected agricultural areas. The optimum production schemes derived *ad hoc* from these computerized applications could offer guidance to the farming units and households in order to manage their production schemes on an overall rational basis.

The conception of the LP scheme in specific agricultural farming units could on the one hand define the productions objectives of desired products and on the other hand could specify the constraints imposed by the necessities of CAPLAM, in terms of quality specifications and financial resources. Some experimental computerized applications in the literature could define the framework of the terms of reference of the activities of farmers in concrete case-studies or selected products.

The European agricultural authorities must determine what we call 'paradigmatic farming units' which will be 'centers of agricultural excellence' for farmers all over Europe according to: (a) geographical-ecological areas, (b) products, (c) size of farming units, (d) quality specification, (e) financial resources capability, and (f) other specifications. A list of 'centers of agriculture excellence' (in terms or economic, ecological, social, cultural and other standards) for European agriculture will rationalize through 'paradigms' the content of CAPLAM in search of the best policy for the benefit of European consumers.

The Content of CAPLAM

The content of CAPLAM is expressed analytically by the «Demeter Oracle Guidelines» which define, in terms of broad objectives, the basic aims of EU agriculture. Demeter was one of the twelve Olympians deities, the Mother Goddess of the Earth, the Patroness and divinity of Agriculture. The following are the «Demeter Oracle Guidelines»:

i) Agriculture is the most vital (literally speaking) human activity since it provides the agricultural products which are of absolute necessity for human life. Since nutrition is the necessity for human life, agriculture deserves the top priority and the strong support of the EU budget resources.

ii) Agricultural production should be oriented towards real consumer needs or market demand as is expressed in equation [1], and should avoid unnecessary surpluses of agricultural products, which constitute a waste of scarce resources.

iii) Emphases and high priority must be given towards the ecological-chemical quality of agricultural products which are produced in the appropriate geographical areas according to their comparative product quality advantage inherent to the specific climatic-ecosystems. Chemical wise, due to different ecological conditions, the quality composition of crops produced in the Mediterranean region is far superior compared to the quality of similar crops which are grown in Central and North Europe. Thus, some crops should be cultivated exclusively in the Mediterranean regions where the climatic and other ecological conditions favor the highest quality of agricultural products for health and the overall nutritional welfare of the European consumers.

iv) All agricultural activities (both crop and livestock production) including the processing and marketing of agricultural products must be carried out in such a way as to fully preserve the quality of the physical environment and the sustainability of all natural resources with particular care taken over the upkeep of the landscape and the countryside.

v) In defining and designing the future of European agriculture emphasis must be given to the wise planning and management of the agricultural affairs rather than to vague «policy measures». In fact, all European farmers should learn and apply the basic principles of management in the operation of their farming units. In this respect, the creation and the promotion of some selected 'centers of excellence for agriculture' could facilitate the paradigmatic and rational implementations of efficient agricultural programs or schemes.

vi) Modern agricultural activity should not be restricted to «monoculture schemes». In order to achieve an increase in 'total agricultural income,' modern farming units should carry out multi-cultural and multi-purpose activities such as arable plus livestock plus processing plus marketing plus agro-business plus agro-tourism, plus reforestation, and other related functions provided that they are planned in a consistent manner so as to raise the overall efficiency of total agricultural activity.

The Basic Principles of CAPLAM

The basic principles or the criteria of the proposed CAPLAM are as follows:

a) The agricultural sector should function on the basis of a need-oriented activity and not a supply-oriented one. This means that total agricultural production by sectors and sub-sectors should be planned according to the needs of the EU as defined by a specific combination of a macro-economic formula, and commodity balances by sectors (according to specific demand functions).

b) The specialization of EU agriculture should be guided on the basis of the ecological-geographic comparative advantage in the Community, that is, production in any country or region should follow the respective 'physical advantage' of each region. (We must not forget that we are planning a predominantly ecologically-based activity). Thus, agricultural production should be oriented to product diversification and quality improvement according to the internal and external market forces. It must also take into consideration the ecological-regional comparative advantage and income elasticities of the individual products. The quality standards of products (product quality) should be defined according to the biochemical analysis of a certified European Food Research Institute. Thus, for example, in the unified European market, if Greek tomatoes are bio-chemically superior to Dutch tomatoes, tomato growing in Holland must eventually be restricted; on the other hand, if milk production in Holland is bio-chemically superior to milk production in Greece, it must be restricted in the latter country. CAPLAM should incorporate most of the EU Regional Development aspects as defined by the Regional Policy.

c) Agricultural production in EU should be planned not in isolation but in the context of the world market for agricultural production as it is 'directed' by the international division of labor and competition. Since prices policy will be 'the principal' instrument in helping to secure the balanced development of European agriculture, this policy should be guided by world prices working as «the shadow prices» in the proposed CAPLAM, taking efficiency into consideration overall.

d) The EU market organizations for various individual products should be reformed according to the above principles. Thus, all intervention mechanisms including co-responsibility levies, public intervention stocks, subsidies etc. should be formed by these principles. Obviously, the stabilization of agricultural markets should be secured through planning and management rather than though policy.

e) The so-called socio-structural policy of the EU dealing with the social aspects of European agriculture should support and help CAPLAM as an integral part of an overall medium and long-term planning process rather than as a disjoined policy approach. The new European agricultural policy must be closely linked to the social characteristics of agricultural activity and should take into consideration that in the agricultural sector economic performance is related to the social way of life of the farming community.

f) The opportunity cost principle should be an important guideline in allocating resources in the EU agricultural economy in the context of CAPLAM. Reallocation of resources should take place among the agricultural sub-sectors (endo-agricultural reallocation) and between the agricultural and non-agricultural sectors (agricultural/industrial balance). Thus, for example, CAPLAM should be the main instrument in transferring resources between the agricultural and industrial sectors of EU aiming at an efficient allocation of resources in the context of an overall agricultural/industrial balance. Some agricultural resources must be planned to be directed towards energy production by biomass (the green energy revolution) while the deterioration of environment by the industrial complexes must be confronted by expanding the appropriate agricultural plantations.

g) Agricultural planning should be comprehensive, including not only production but also product disposal, by organizing the processing and marketing activities of EU agriculture. However, the new European agricultural policy must respect the cultural heritage of the agricultural way of life. Preserving the cultural heritage of the European farming community must be an important criterion in shaping CAPLAM.

h) Any planning activity in EU agriculture should also take into consideration the economic and the socio-cultural status of the agricultural household as the basic unit of CAPLAM. In fact, planning will be successful only if the family farm (the agricultural household) or the farming firm become an integral part of CAPLAM.

i) Improving European agricultural technology (in selected areas), by directing more resources to that sector should be an important part of CAPLAM. Thus, *ad hoc* plans (specific projects) should be financed to promote R&D in the European agriculture, so as to improve its quality standards and its impact on sustainability.

The Macroeconomic Aspects of CAPLAM

The planning of agricultural production and disposal in EU should satisfy the following macroeconomic formula which defines total planned supply and its allocation to various uses:

$$PA + PM = C + X + S + BE + RC + WF + L \qquad [5]$$

that is:

- on the supply side:
  PA = total planned agricultural output in EU, and
  PM = planned or estimated imports from the non-EU world
- on the demand side (uses):

C = need satisfaction defined as food consumption plus food processing in the EU,
X = planned or estimated exports to the non-EU world (European agricultural exports could increase far beyond the present level),
S = normal stocks (approximately 2% of PA),
BE = a provision for biomass energy resources (being a growing percentage as a result of the gradual substitution of exhaustible energy resources to renewable agricultural energy resources),
RC = a provision for the reproductive cycle of agriculture, that is, seeds + animal feeding and upgrading + provision for anti-erosion + environmental protection,
WF = international food aid, that is, a provision for the nutrition of the third world poor population as referred to earlier (over 2% of the PA), and,
L = a provision for normal losses of perishable agricultural production.

It is important to note that planned agricultural production (*PA*) in EU should provide not only for food consumption in the EU but for the nutrition of some of the world hungry populations (*WF*). In fact, most of the aid to the less developed world should take the form of the distribution of some of the food surpluses to the hungry people of the world. It is also desirable that items *BE* and *RC* in [5] should absorb a greater part of the EU agricultural resources in the long-run. These items are virtually ignored in the existing CAP; CAPLAM should take care to incorporate them in its more comprehensive framework.

The Opportunity Cost of CAPLAM

The most appropriate strategy of dealing with the EU «agricultural problem» is to plan the transfer of some mobile resources from the agricultural sector to the other sectors where these resources could be more efficiently employed, thus ensuring a resource allocation balance in the European economy. In fact, the opportunity cost of using excess resources now in EU agriculture seems to be high since a great part of these resources could be used in other more productive activities, provided that the transfer of these resources is politically and socially feasible. Planning is more efficient in fighting the political and social constraints in this respect compared to policy maneuvering and tactics.

After estimating the excess resources in European agriculture, the proposed planning in EU agriculture, that is CAPLAM, could undertake selective and progressive action to ensure the transfer of surplus agricultural resources to the sectors where they would enjoy a higher social and private return. This difficult exercise could be carried out by planned action rather than by spasmodic policy measures (the superiority of the «planning method» compared to the «policy method»). Thus, it is hypothesized that the existing surplus resources in EU agriculture could be reduced to an optimum level which is necessary to produce *PA* in formula [5] and substitute some of the imported agricultural goods (*PM*). Obviously, it is assumed that there is a optimum level of resources in EU agriculture (*ORA*) and that the existing resources devoted to agricultural production (*RA*) are in excess of that optimum level:

$$RA > ORA \qquad [6]$$

Therefore, the surplus of resources now engaged in European agriculture (*SRA*) is defined as follows:

$$SRA = RA - ORA \qquad [7]$$

One of the final objectives of agricultural planning in the EU (CAPLAM) is to ensure the transfer of *SRA* to the other sectors, based on allocative efficiency criteria. Some macro-linear-programming models could be used to estimate *RA* and *SRA* and the allocation of *SRA* to the non-agricultural sectors in order to approach the agricultural/industrial balance in the EU economy.

Planning at the Macro, Sectoral, and Micro Levels

The long-term solution to the EU 'agricultural problem' and the achievement of the agricultural/industrial balance in the EU necessitates the use of wise planning at the macro, sectoral and micro levels. Following the basic principles as explained already, a selective and progressive planned action should be undertaken to rationalize the existing agricultural economy in the EU. The emphasis should be mainly on micro-planning, that is, planning at the family farm level or at the individual farm business level since these are the ultimate decision makers in shaping EU agriculture based on CAPLAM prescriptions. There are a lot of plans, adjustments, measures etc. at the farm level which could be used to cope with the agricultural problems of EU in a more efficient way. In its present form the CAP relies mainly on a «general institutional approach» which is not so efficient (marathon meetings of the Agricultural Ministers, price negotiations etc.). The proposed CAPLAM is designed to deal with agriculture both in the context of the whole economy (agricultural/industrial balance) and at the family farm level (farm budgets), trying to manage European agriculture both from «top down» and «from below».

EU farming organizations should take an active role in shaping CAPLAM since the objective of this approach is to maximize the (gross and net) income of the farmers by exploiting their idle surplus resources to the benefit of all Europeans since this method will eventually achieve the agricultural/industrial balance in the EU.

The important weakness of CAP is that it relies mainly on pricing and subsidies policies. CAPLAM would be designed to synthesize the interaction between resource allocation, production, consumption, pricing, stocking, processing, marketing, and the international aspects of EU agriculture in an overall planning and management scheme. Nothing should be designed without further research at the macro, sectoral and micro levels. Allocative and distributive efficiencies in the planning context would be established using farm management, linear programming, cost-benefit analysis, farm budget analysis and other techniques. A comprehensive research agenda would define the content and the limitations of CAPLAM.

Implementing CAPLAM

There are some crucial factors which influence the efficient implementation of a plan (implementation efficiency). Among these factors are the following: the quality of decision-makers and administrators; the quantity of resources devoted to the implementation process; the feasibility of the desired objectives; the existing political and social constraints; the degree of mobilization of public opinion. The overall planning and management of CAPLAM must systematize the above factors in order to achieve the best results compared to the CAP.

The Ministries of Agriculture of the respective member countries of the EU could act as decentralized agencies of a Central Agricultural Planning Commission in Brussels. This would control the performance of CAPLAM, using efficient implementation techniques (guided targets, incentives, disincentives, and other market-oriented planning instruments).

The existing CAP is highly centralized in Brussels and remote from the exact area problems at the local level. The proposed CAPLAM would try to synthesize central planning in Brussels with new peripheral-regional planning agencies of the EU attached to the Ministries of Agriculture in each region of member countries promoting the regional implementation of CAPLAM for the real benefit of European agriculture. Problems of agricultural competitiveness and sustainability in the EU would be greatly facilitated by adopting this proposed new strategy.

Bibiography

Antle, J. M., J. N. Lekakis and G. P. Zanias (eds), 1998. *Agriculture , Trade, and Environment: The Impact of Liberalization on Sustainable Development*, Cheltenham: Edward Elgar (forthcoming).

Avery, G. 1984. «Europe's Agricultural Policy,» *International Affairs*, 60(4): 643-656.

Commission of European Communities, 1991. *The Development and Future of the Common Agricultural Policy*, COM (91) 258 final/3.

Commission of the European Communities, 1992. *Towards Sustainability: A European Community Programme of Policy and Action in Relation to the Environment and Sustainable Development*, Brussels.

European Community, 1991. *Evolution et Avenir de la Politique Agricole Commune*, Communication de la Commission an Conseil et Parlement Européen, COM (91) 258 final, Bruxelles.

European Community, 1995. «Agricultural Paper,» Communication by M. Fisher to Council, Brussels.

Goldin, I. and O. Knudsen, 1990. *Agricultural Trade Liberalization: Implications for Developing Countries*, OECD, Paris.

Guyomard, H. and L. P. Mahé, 1995. «La Nouvelle Instrumentation de la Politique Agricole Commune,» *Economie et Prevision*, 117-18: .15-29.

Hathaway, D. 1987. «Agriculture and the GATT: Rewriting the Rules, Policy Analyses in International Economics", no. 20, Washington: Institute for International Economics.

Johnson, M., L. P. Mahé, and T. L. Roe, 1993. «Trade Compromises between the European Community and the United States: An Interest Group-Game Theory Approach,» *Journal of Policy Modelling*, 15: 199-222.

Krueger, A., M. Schiff, and A. Valde, 1988. «Agricultural Incentives in Developing Countries: Measuring the Effect of Sectoral and Economywide Policies,» *The World Bank Economic Review*, 2-3: 255-271.

Mahé, L. P. and T. L.Roe, 1996. «The Political Economy of Reforming the 1992 CAP Reform,» *American Journal of Agricultural Economics*, 78: 1314-1323.

OECD, 1995. *The Uruguay Round: A Preliminary Evaluation of the Impacts of the Agreement on Agriculture in the OECD Countries*, Paris.

Parikh, K. S., G. Fisher, K. Frohberg, and O. Gulbrandsen, 1987. *Towards Free Trade in Agriculture*, Amsterdam: Nijhoff.

Rausser, G.C. and D.A. Irwin, 1989. «The Political Economy of Agricultural Policy,» *European Review of Agricultural Economics*, 15: 349-366.

Redclift, M. R. , J. N. Lekakis, and G. P. Zanias (eds), 1998, *Agriculture and World Trade Liberalization: Socio-environmental Perspectives on the Common Agricultural Policy*, Wallingford: CAB International (forthcoming).

Theofanides, S. M. 1979. «The Static Economic Effects of Greece's Entry into the EEC: The Case of Olive Oil–A Cost-Benefit Analysis,» *European Review of Agricultural Economics*, 6(1): 81-90.

Theofanides, S. M. 1982. «Key Issues in Project Planning and Implementation: The Case of Greece,» in W. D. Cook and T. E. Kuhn (eds), *Planning Progress in Developing Countries: Techniques and Achievements*, TMS Studies in the Management Sciences, 17: 47-87, Amsterdam: North-Holland.

Theofanides, S. M. 1987. «European Growth: Some Mega-Scenarios on the Distribution Aspects of Gains and Losses,» paper presented at the Third Conference of the "Europe-12" Committee, 18-19 September, Bonn University.

Theofanides, S. M. 1992. «Improving the Institutional Framework of the European Community,» in *Studies in Honor of Professor A. Kanellopoulos*, Vol. A΄, University of Piraeus, Piraeus.

Urff, W. V. 1984. «Die gemeinsame Agrarpolitic-Integrationsopfer der Deutschen?» in R. Hrbek und W. Wessels (eds), *EG-Mitgliedschaft: ein vitales Interesse der Bundesrepublik Deuschland*, Bonn.

Xenophon, *Oikonomikos*, Boston: Harvard University Press, 1969 (The first interdisciplinary textbook in Agricultural Economics; an inaccurate translation in English from the Greek original written in 394 B.C.).

Author Index

Subject Index

Economy & Environment

1. F. Archibugi and P. Nijkamp (eds.): *Economy and Ecology: Towards Sustainable Development.* 1989 ISBN 0-7923-0477-2
2. J. Bojö, K.-G. Mäler and L. Unemo: *Environment and Development: An Economic Approach.* 1990 ISBN 0-7923-0802-6
3. J. B. Opschoor and D. W. Pearce (eds.): *Persistent Pollutants: Economics and Policy.* 1991 ISBN 0-7923-1168-X
4. D.J. Kraan and R. J. in 't Veld (eds.): *Environmental Protection: Public or Private Choice.* 1991 ISBN 0-7923-1333-X
5. J.J. Krabbe and W.J.M. Heijman (eds.): *National Income and Nature: Externalities, Growth and Steady State.* 1992 ISBN 0-7923-1529-4
6. J. Bojö, K.-G. Mäler and L. Unemo: *Environment and Development: An Economic Approach* (revised edition). 1992 ISBN 0-7923-1878-1
7. T. Sterner (ed.): *Economic Policies for Sustainable Development.* 1994 ISBN 0-7923-2680-6
8. L. Bergman and D.M. Pugh (eds.): *Environmental Toxicology, Economics and Institutions.* The Atrazine Case Study. 1994 ISBN 0-7923-2986-4
9. G. Klaassen and F.R. Førsund (eds.): *Economic Instruments for Air Pollution Control.* 1994 ISBN 0-7923-3151-6
10. K. Uno: *Environmental Options: Accounting for Sustainability.* 1995 ISBN 0-7923-3513-9
11. K. Uno and P. Bartelmus (eds.): *Environmental Accounting in Theory and Practice.* 1997 ISBN 0-7923-4559-2
12. J.C.J.M. van den Bergh, K.J. Button, P. Nijkamp and G.C. Pepping: *Meta-Analysis in Environmental Economics.* 1997 ISBN 0-7923-4592-4
13. S. Faucheux, M. O'Connor and J. v.d. Straaten: *Sustainable Development: Concepts, Rationalities and Strategies.* 1998 ISBN 0-7923-4884-2
14. P. Kågeson: *Growth versus the Environment: Is there a Trade-off?.* 1998 ISBN 0-7923-4926-1
15. J.C.J.M. van den Bergh and M.W. Hofkes (eds.): *Theory and Implementation of Economic Models for Sustainable Development.* 1998 ISBN 0-7923-4998-9
16. J.N. Lekakis (ed.): *Freer Trade, Sustainability, and the Primary Production Sector in the Southern EU: Unraveling the Evidence from Greece.* 1998 ISBN 0-7923-5151-7

KLUWER ACADEMIC PUBLISHERS – DORDRECHT / BOSTON / LONDON

GPSR Compliance

The European Union's (EU) General Product Safety Regulation (GPSR) is a set of rules that requires consumer products to be safe and our obligations to ensure this.

If you have any concerns about our products, you can contact us on ProductSafety@springernature.com

In case Publisher is established outside the EU, the EU authorized representative is:

Springer Nature Customer Service Center GmbH
Europaplatz 3
69115 Heidelberg, Germany

Batch number: 10823959

Printed by Printforce, the Netherlands